Grasses and Grassland - New Perspectives

Edited by Muhammad Aamir Iqbal

Published in London, United Kingdom

IntechOpen

Supporting open minds since 2005

Grasses and Grassland - New Perspectives
http://dx.doi.org/10.5772/intechopen.95204
Edited by Muhammad Aamir Iqbal

Contributors

Ang Jia Wei Germaine, Sanjay Kumar Sanadya, Surendra Singh Shekhawat, Smrutishree Sahoo, Marcelo de Andrade Ferreira, Luciano Patto Novaes, Michelle Christina Bernardo de Siqueira, Ana Maria Herrera Ângulo, Tessema Tesfaye Atumo, Bereket Zeleke Tunkala, Milkias Fanta Heliso, Derebe Kassa Hibebo, Yoseph Mekasha, Greg Aldrich, Renan Donadelli, Pius Yoram Kavana, John Kija Bukombe, Hamza Kija, Stephen Nindi, Ally Nkwabi, Iddi Lipende, Sood Ndimuligo, Simula Maijo, Baraka Naftali, Victor M. Kakengi, Janemary Ntalwila, Robert Fyumagwa, YanQing A. Zhang, Neil West, Muhammad Aamir Iqbal, Sadaf Khalid, Raees Ahmed, Muhammad Zubair Khan, Nagina Rafique, Raina Ijaz, Saira Ishaq, Muhammad Jamil, Aqeel Ahmad, Amjad Shahzad Gondal, Muhammad Imran, Junaid Rahim, Umar Ayaz Aslam Sheikh

Notice

Statements and opinions expressed in the chapters are these of the individual contributors and not necessarily those of the editors or publisher. No responsibility is accepted for the accuracy of information contained in the published chapters. The publisher assumes no responsibility for any damage or injury to persons or property arising out of the use of any materials, instructions, methods or ideas contained in the book.

First published in London, United Kingdom, 2022 by IntechOpen
IntechOpen is the global imprint of INTECHOPEN LIMITED, registered in England and Wales, registration number: 11086078, 5 Princes Gate Court, London, SW7 2QJ, United Kingdom
Printed in Croatia

British Library Cataloguing-in-Publication Data
A catalogue record for this book is available from the British Library

Additional hard and PDF copies can be obtained from orders@intechopen.com

Grasses and Grassland - New Perspectives
Edited by Muhammad Aamir Iqbal
p. cm.
Print ISBN 978-1-83969-833-0
Online ISBN 978-1-83969-834-7
eBook (PDF) ISBN 978-1-83969-835-4

Meet the editor

Dr. Muhammad Aamir Iqbal is a Pakistani researcher with expertise in the production of cereal–legume forages, preservation of forages, intercropping systems, and grassland management. He obtained a Ph.D. in Agronomy from the University of Agriculture Faisalabad, Pakistan, and underwent research training at the University of California Davis, USA, and the Shandong Academy of Agricultural Machinery Sciences, China. He has more than a decade of research and teaching experience. Dr. Iqbal has more than 103 research and review articles as well as numerous book chapters to his credit and is an editor and reviewer of many reputed journals. Presently, Dr. Iqbal is an assistant professor in the Department of Agronomy, Faculty of Agriculture, University of Poonch Rawalakot, Pakistan.

Contents

Preface

We are in an era of changing climate, global warming, disturbed rainfall patterns, droughts, and floods, all of which are major stresses on food production systems. This situation demands devising and implementing novel solutions to boost productivity and farmers' profitability as well as restore and preserve terrestrial ecosystems. *Grasses and Grassland - New Perspectives* discusses grassland management, utilization, and restoration under the changing climate. It also includes vital information on potential uses and future perspectives of many grass species.

This book includes two sections. The first section, "Trends in Grassland Management," presents fundamental concepts, recent knowledge, and advancements in the management, economic utilization, effective restoration, and viable preservation of grasslands. The second section, "Underutilized Grasses Production Potential," discusses grasses that have remained neglected despite their potential to ensure food security for the skyrocketing population under changing climate and global warming.

Chapter 1 provides fundamental knowledge pertaining to the concepts of grasslands (prairie, savanna, steppe, pampas, etc.), different types of grasslands (natural, semi-natural or improved, tropical, temperate, tundra, montane, xeric, and flooded grasslands) and elaborates different ecosystem services (processes, conditions, and outputs) provided by grasslands. These services include being a source of feed for ruminants, serving as habitats for ensuring species biodiversity, mitigating drought, flood, and soil erosion, and much more. It elucidates the reasons why grasslands need to be developed to support ecosystem services provided by fragile existing grasslands. Finally, the chapter suggests developing grasslands for generating a green economy via the adoption of integrated approaches encompassing integrated fertilization regimes, over-seeding of leguminous plant species, adjusting herbage allowance, manipulating stocking rate, and monitoring using global positioning systems and infrared spectroscopy.

Chapter 2 discusses ecosystem services provided by grasslands in terms of generating livestock-rearing opportunities and thereby increasing the livelihood of stakeholders. It highlights grassland ecosystem services with respect to sustaining the wildlife in Tanzania. It also suggests that new research is needed to restore and preserve Tanzanian grasslands.

Chapter 3 examines recent developments in climate change and emerging stresses, especially abiotic stresses like drought, and their role in grass diseases. It provides information on common diseases such as *Xanthomonas* spp. and *Pseudomonas* spp. and their interaction with vectors. The chapter also suggests strategies to break the cycle of vector growth, regrowth, and infestation.

The second section begins with Chapter 4 on the production of underutilized grasses and their potential in temperate and tropical regions. The chapter elaborates on highly debated aspects pertaining to underutilized grasses and neglected grasses and distinguishes those in an explicit manner. Additionally, it highlights underutilized grasses' potential as food, feed, fuel, energy crops, and medicinal purposes. Lastly, it shares vital information on potential strategies to boost production and applications of underutilized grasses.

Chapter 5 presents information on the potential of miscanthus grass to provide nutritional fiber to monogastric animals. It discusses miscanthus grass in pet feed as well as provides suggestions to improve chicken and pet health by optimizing their feed fiber content.

Chapter 6 discusses the effectiveness of plant nutrition management strategies on grasses productivity in terms of biomass yield. It shows the comparative efficacy of different doses for boosting herbage yield and marginal rate of return.

Chapter 7 examines the ecosystem classification system in Western Utah and the Yukon territory in Canada.

Chapter 8 discusses how to address the forage shortage in semi-arid regions, focusing on the spineless cactus, which is a perennial crop. It also examines the productivity potential of cactus in Brazil and production technology to boost biomass yield in semi-arid conditions.

Finally, Chapter 9 discusses the potential of Sewan grass to be grown as a forage crop in arid regions as well as its production and distribution.

There are not enough words to express due gratitude to Almighty Allah (the sustaining source of kindness) whose mercies and exaltation enabled me to take on the task of serving as editor of this book, which required the utmost commitment and dedication. All wisdom and intellect belong to Him and it was His countless blessings that helped me along the way in compiling this book. All praises and compliments for Prophet Muhammad (O Allah! Send Your Mercy on Muhammad and on the family of Muhammad, as You sent Your Mercy on Abraham and the family of Abraham, for You are the Most Praise-worthy, the Most Glorious), Who is ultimate educator and ever-lasting source of knowledge for whole humanity. In addition, I would like to acknowledge the strategically vital and pertinent intellectual support furnished by our mentor Dr. Asif Iqbal and team members, including Dr. Raees Ahmed, Dr. Muhammad Imran, Dr. Junaid Rahim, Dr. Umer Ayaz Aslam Sheikh, Dr. Muhammad Jamil, Mrs. Sadaf Khalid, and Dr. Bilal Ahmad.

Key Features of the book:

- Furnishes fundamental and state-of-the-art knowledge on grasslands management and ecosystem services.

- Provides vital information on different underutilized grasses and their potential as food, fuel, beverages, and energy crops.

- Presents integrated strategies to boost restoration and production of grasslands for ensuring food security of future generations.

- Describes and discusses different classes of grasslands.

- Illustrates potential strategies to boost the utilization of underutilized grasses.

- Highlights the importance of grasses for attaining the goals of poverty alleviation and zero hunger.

- Depicts global diversity by sharing knowledge from a community of international researchers.

Muhammad Aamir Iqbal
Faculty of Agriculture,
Department of Agronomy,
University of Poonch Rawalakot,
Azad Jammu & Kashmir, Pakistan

Trends in Grassland Management

Introductory Chapter: Grasslands Development - Green Ecological Economy and Ecosystem Services Perspectives

Muhammad Aamir Iqbal

1. Introduction

Globally, grasslands known by the names of prairie, savanna, steppe, and pampas in conjunction with rangeland occupy over 70% of the agricultural area of which 68% lies in the developing countries. Grasslands provide a variety of foods and forages while people also rely heavily on them for their source of earning through milk, meat, and wool production. Over time, more than 20% of the world's native grasslands have been transformed into croplands to carry out intensive farming of cash crops. There are over 1 billion of the world's poorest people depend on the livestock industry, which relies on native grasslands for animal feed. In this way, grasslands support the production of over one-third of protein requirements worldwide [1–10].

In many developed countries of Europe and North America, the native grasslands have been continuously converted into pasturelands for boosting milk production or croplands for cultivating high-yielding grain and cash crops. The extent of grasslands transformation might be realized from the fact that tall-grass prairie spreading across many states of the US has been converted to carry on intensive farming of crops, leaving behind less than 1% of the original prairie. Contrastingly, many developing countries of Africa and Asia have kept on extensively utilizing their native grasslands as a source of cost-effective feed source and watershed.

2. What are grasslands?

Grasslands constitute one of the primary and largest biomasses on earth which dominate all types of natural landscapes on all habitat-able continents of the world except Antarctica. In simplest words, grasslands may be defined as areas whereby the most dominant vegetation are grasses belonging to the family *Poaceae*, however, other flora such as various types of sedges of *Cyperaceae* family along with different rushes classified in *Juncaceae* family can also constitute a minor chunk of local eco-region. Additionally, grasslands being the habitat of biodiversity (flora and fauna) may also contain variable proportions of legumes species belonging to *Fabaceae* (*Leguminosae*) and various other herbs. Grasslands have also been described as terrestrial ecosystems, which are dominated by various herbaceous vegetation and different kinds of shrubs whereby plant species biodiversity gets regulated and maintained by factors such as grazing intensity, fire, grazing, and temperatures (scorching and chilling), rainfall intensity and distribution, etc. Furthermore, semi-natural grasslands are formed owing to human activities (mowing and grazing), while environmental growth conditions (temperature, precipitation, relative humidity,) and natural processes such as fire, floods, drought, etc. determine the species pool and genetic diversity of grass species [11].

3. Classification of grasslands

Unimproved grasslands are dominated by unsown plant species and wild vegetation communities and can be either natural (having no planned grazing or mowing, over-seeding, etc.) or semi-natural (natural plant communities such as grasses, sedges, rushes, and herbs that are maintained by anthropogenic activities including grazing and planned biomass harvesting regimes) grasslands.

Another type of major grassland is tropical grasslands that are situated around the equator (between the tropic of Cancer and Capricorn) in the interior of continents. These serve as a point of segregation between rainforests and deserts. These are also known by the name of Savannahs. These witness tropical continental climates and have alternate wet and dry seasons. Examples include hot savannas of sub-Saharan Africa and the northern grasslands of Australia (called rangelands). In contrast, temperate grasslands are found in the north of the tropic of Cancer and south of the Tropic of Capricorn. These grasslands have a cooler climate compared to Savannahs, which is called temperate continental climate. Examples include North American prairies, Eurasian steppes, and Argentine pampas. In addition, tundra grasslands also referred to as polar grasslands are located in higher altitudes in subarctic regions having a very short vegetation growing season. Furthermore, the grassland found above the tree-line at high altitudes is commonly known by the name montane (literal meaning of high altitude) grasslands. The plant species of these grasslands are quite unique in the agro-botanical structure having specific dish-like formation along with the presence of thick waxy surface plant area. A typical example of montane grasslands include Northern Andes. Moreover, xeric grasslands, also called desert grasslands, are located around the desert low lands. Lastly, flooded grasslands tend to have abundant water throughout the year and contain a variety of vegetation that thrives well in water. Numerous types of water birds frequently migrate to flooded grasslands, while a typical example includes the everglades grassland, which is referred to as the world's largest flooded grassland [12].

4. Why grasslands development needed?

Grasslands development occupies a pivotal position keeping in view the fact that these are located in regions wherein their rainfall is insufficient to effectively support the growth of trees to form a rain trees forest, but not so scarce to form a desert. Thus, it may be inferred that grasslands often serve as a transition zone between deserts and forests. These serve as one of the prime ecosystems and cover over one-third of the terrestrial surface worldwide. Extensively managed grasslands have emerged as one of the most secure habitats to ensure plant biodiversity. The need for their development even becomes more important as grasslands in conjunction with different rangelands contribute significantly to boost livestock productivity through the provision of cost-effective and nutritious forage abundantly and that too throughout the year owing to grasses diversity containing perennial grass species. Another aspect emphasizes the pertinence of grasslands development as grasslands (both natural and semi-natural grasslands) play a vital role in the provision of life-sustaining livelihood to people by providing animal feed. Developing grasslands has become mandatory, keeping in view the rapidly increasing supply needs for animal products owing to skyrocketing human population. In addition, gradually hiking consumption patterns and demand for livestock products (milk, meat, wool, etc.) on per capita basis has made it necessary to increase the conversion of natural grasslands into improved grasslands. It should be kept in mind that competition and land-use patterns are predicted to multiply considerably by 2050, which may be accentuated by the recent scenario of climate change. This scenario increased the intense focus on sustainable food production for ensuring food security through alteration of agricultural sciences research approaches and policymaking at state, regional, and global levels. Grasslands development can be achieved by putting into use the sustainable intensification concept, in terms of increasing the productivity of grasslands in order to supplement the production potential of croplands. However, up till now, the role of improved grasslands through biologically viable improvement and development has been direly neglected and thus compromising the food security of many tropical grassland regions of Africa and Asia [11–14].

The developed grasslands might be of unprecedented pertinence due to having very high conservation value and the potential to support sustainable food production. The co-development of grasslands adjacent with various types of rangelands, including shrubland and savannas can contribute significantly in ensuring the survival and food security of the surrounding population. Grassland development has to be initiated by keeping in view their local importance in terms of ensuring and maintaining the species biodiversity as well as food production. In addition, these also influence a variety of ecological processes at the local landscape (pollination), regional level (water regulation and recreation activities), and global scales (climate regulation) which necessitate their development in an integrated manner without disruption of prevalent ecosystems. Grasslands provide feed base to grazing livestock for producing high-quality food products, and in return get organic manures, a source of pollination and planting material transportation through natural means along with the provision of leather for human utilization for various purposes. In addition, grassland development can potentially provide vital services and roles such as water catchments, reserves of biodiversity, and fulfilling cultural as well as recreational needs. More importantly, grassland development has the potential to increase their capacity to serve as a carbon sink for alleviating the emissions of greenhouse gases which have contributed heavily to global warming and climate change. Inevitably, grasslands development might invoke plenty of challenges, but those have to be confronted and tackled through target-oriented and collaborative research and policymaking in a coherent manner.

5. Grasslands development strategies and green ecological economy

The sustainable development of grasslands requires the adoption of integrated approaches for ensuring the grasslands improvement, having minimum disruption of local ecological systems and non-significant adverse effects on biodiversity of plant species [1, 3, 15, 16]. Different biologically viable strategies for grasslands development may include fertilizer application keeping in view the optimal combination of chemical fertilizers and organic manures along with planned grazing management. In addition, boosting the use of crop by-products such as green compost application for increasing grasslands soil fertility status, over-seeding of native leguminous plant species and manipulation of stocking rate (animal numbers that can be successfully reared on a specific land area over a certain time period and expressed as animal units per unit land area) might be used as effective strategies for grasslands development. In addition, herbage allowance (grams of herbage dry matter per kg live weight per day per animal unit) adjustment offers one of the feasible solutions to over-grazing and over-utilization of grasslands.

To the best of our knowledge, concrete findings based on empirical results are still lacking for estimating and predicting the utilization efficacy and cost-effectiveness of grasslands development strategies. The situation is even worse for grasslands production systems and the instance of grasslands in sub-Himalayan regions of Jammu and Kashmir can be taken as a gauge study. The scientific evaluation and appropriate management of prevalent grazing systems need reliable and feasible assessment criteria without which grasslands productivity improvement will continue to remain a distant dream. Recently, a bunch of emerging technologies has contributed significantly in acquiring the timely and low-cost quantitative information for understanding the complex soil-pasture-grazing animals' interactions along with animal management with respect to grassland systems capacity and potential under changing climatic scenario. For instance, remote imaging might be useful for estimating the vegetation status in particular inaccessible grassland. In addition, a global positioning system (GPS) can also be put into practice for monitoring natural or man-induced factors like fire and over-seeding requirements due to heavy and uncontrolled grazing in a specific patch(s) of natural or improved grasslands. Moreover, improved diet markers and near-infrared (IR) spectroscopy along with using different modeling techniques may provide concrete and real-time information in order to take knowledge-based decisions regarding productivity constraints of grasslands and grazing animals. Furthermore, using individual electronic identification (EI) of different grazing animals may offer unprecedented opportunities to go for precision management of animal units that is bound to improve the productivity of milch animal, especially large ruminants. However, it must be noted that sustainably better and improved outcomes in terms of grazing animal products, services, and various by-products from natural or improved grasslands, can be feasible depending on devising clear and viable solutions that can be successfully employed in diversified environments and socio-technological circumstances of grasslands managers globally.

6. Grasslands and ecosystem services

In simplest words, ecosystem services are defined as "various outputs, conditions, and processes of natural biological systems which in one way or other, directly or indirectly, benefit humans and significantly enhance their social welfare" [13, 16, 17]. These include a variety of processes by which grasslands produce a bunch of beneficial resources including forage for ruminants, clean water by serving as excellent watershed, ensuring biodiversity by offering favorable habitat to wildlife, etc. Globally, extensively managed grasslands have been recognized for having very high biodiversity that assists in maintaining and promoting a variety of social and cultural norms and values. Cultivated grasslands provide the maximum herbage yields of nutritious green forage for feeding grazing animals and various other benefits as illustrated in **Figure 1**. However, the range of ecosystem services offered by them is on the lower side compared to permanent grasslands in terms of total biomass production, herbs biodiversity for preparing cosmetics, etc.

In contrast to cultivated grasslands, permanent ones tend to provide a wider variety of ecosystem services as depicted in **Figure 2**. These grasslands maintain higher diversity of plant and animal species along with providing abundant herbs

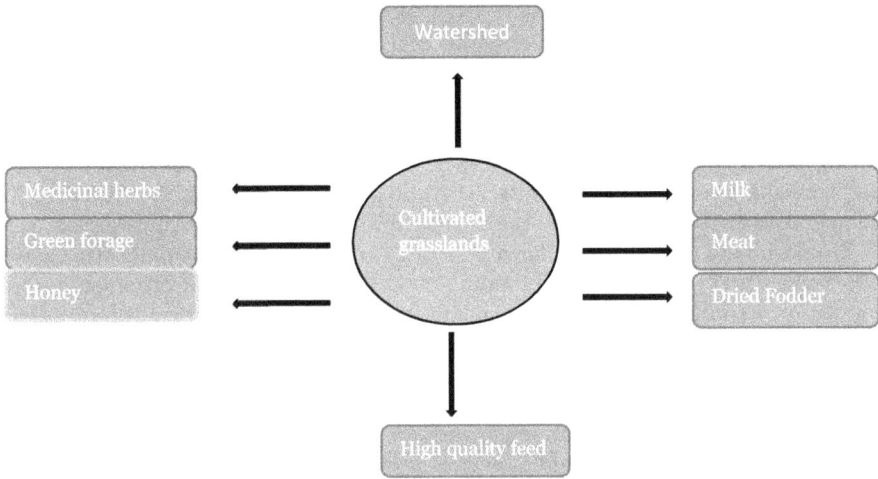

Figure 1.
Various types of ecosystem services are offered by cultivated grasslands under changing climate scenarios.

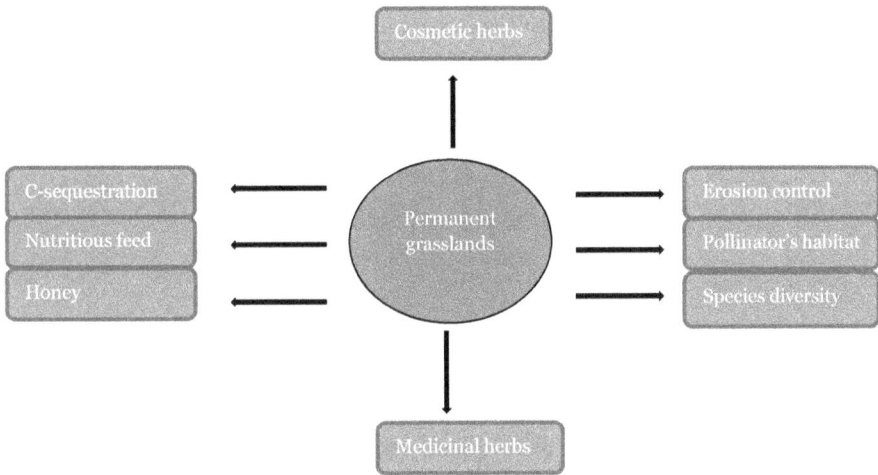

Figure 2.
Various types of ecosystem services are offered by permanent grasslands under changing climate scenario.

for medicine and cosmetics preparation and honey. However, biomass production is significantly lesser than in cultivated or improved grasslands and resultantly grazing ruminant's productivity is comparatively suboptimal. Lastly, semi-natural grasslands tend to have mixed characteristics of improved and natural grasslands, such as higher species diversity coupled with maximum nutrition biomass production owing to fertile soils.

Besides aforestated ecosystem services, grasslands tend to offer many other benefits such as seeds dispersal and preservation of abundant and endangered plant species, flood, and drought mitigate through effective maintenance of microclimate, recycling of macro and micronutrients in the soil as the plant life cycle goes on and detoxification of different wastes through decomposition [18, 19]. Additionally, grasslands ensure species biodiversity by providing suitable habitats and significantly contribute stability to micro and macro climate by restoring natural processes. Furthermore, these serve as an effective source to keep pests under threshold levels due to higher biodiversity which maintains the predator-prey relationship in a natural way. Moreover, protecting the grasslands soil from different types of erosion (sheet and gully erosion) by maintaining living mulch or cover is one of the vital benefits offered by grasslands, which leads to the provision of clean water through protected watersheds. Lastly, the provision of recreation facilities owing to natural or improved esthetic value along with serving as excellent wetland and furnishing research opportunities (natural grasslands and cultivated lands comparative analyzes) are few of the ecosystem's services offered by grasslands.

In addition to agricultural-related benefits, grasslands can potentially offer some other benefits as well such as maintaining water supply and regulation of water flow regulation, carbon sequestration, mitigation of climate, and cultural advantages. To conclude, three types of ES can be extracted from grasslands including animals related ES services (nutritious forage production), cultural (recreation purpose), and micro-environment regulating ES services (pollination, biological control of different insect-pest, mitigation of gaseous emissions). There exist multiple synergies and trade-offs among ES services provided by grasslands and prevalent management practices, however appropriate management practices may potentially create more synergies and reduce trade-offs leading to the sustainable improvement of ES services. It is suggested that grasslands ES services and food security research along with policymaking must be given higher priority for boosting ruminant productivity alongside other ES services. A vital approach that integrates grasslands with modern agricultural production systems as well as land-use patterns optimization at the local and regional level can significantly improve livelihoods and food security. However, future research must focus on grasslands capacity to deliver a variety of ES services in relation to agricultural systems in order to develop sustainable, biologically viable, and economically attractive management options and strategies.

7. Conclusions

Different types of grasslands in conjunction with rangelands occupy over 70% of the agricultural area of which 68% lies in the developing countries whereby their rapid conversion to croplands remains unabated. The deterioration of grasslands may compromise the provision of ecosystem services such as food and feed availability, wildlife habitat disruption, decline in species biodiversity, increase in the number of endangered species, and enhancement of greenhouse gaseous emission owing to lesser C-sequestration. Thus, scientific development of grasslands through optimized management practices that integrate agronomic approaches (appropriate fertilization and balanced over-seeding) with planned utilization (through stocking rate and herbage allowance adjustment) and real-time monitoring using the latest techniques (GPS and IR spectroscopy) hold the potential to offer compatible benefits leading to improved productivity and halting grasslands conversion to croplands. The optimized implementation of integrated management approaches can turn grasslands into green ecological economies offering numerous advantages such as improved livelihood through enhanced milk, meat, wool, and honey production, climate mitigation, control of floods and droughts, watershed management, and wildlife conservation.

Author details

Muhammad Aamir Iqbal
Faculty of Agriculture, Department of Agronomy, University of Poonch Rawalakot, Pakistan

†Address all correspondence to: aamir1801@yahoo.com

IntechOpen

References

[1] Kizilgeci F, Yildirim M, Islam MS, Ratnasekera D, Iqbal MA, Sabagh AE. Normalized difference vegetation index and chlorophyll content for precision nitrogen management in durum wheat cultivars under semi-arid conditions. Sustainability. 2021;**13**:3725

[2] Abbas RN, Arshad MA, Iqbal A, Iqbal MA, Imran M, Raza A, et al. Weeds spectrum, productivity and land-use efficiency in maize-gram intercropping systems under semi-arid environment. Agronomy. 2021;**11**:1615

[3] Haque MM, Datta J, Ahmed T, Ehsanullah M, Karim MN, Akter MS, et al. Organic amendments boost soil fertility and rice productivity and reduce methane emissions from paddy fields under sub-tropical conditions. Sustainability. 2021;**13**:3103

[4] Chowdhury MK, Hasan MA, Bahadur MM, Islam MR, Hakim MA, Iqbal MA, et al. Evaluation of drought tolerance of some wheat (*Triticum aestivum* L.) genotypes through phenology, growth, and physiological indices. Agronomy. 2021;**11**:1792

[5] Iqbal MA, Rahim J, Naeem W, Hassan S, Khattab Y, Sabagh A. Rainfed winter wheat (*Triticum aestivum* L.) cultivars respond differently to integrated fertilization in Pakistan. Fresenius Environmental Bulletin. 2021;**30**(4):3115-3121

[6] Alghawry A, Yazar A, Unlu M, Çolak YB, Iqbal MA, Barutcular C, et al. Irrigation rationalization boosts wheat (*Triticum aestivum* L.) yield and reduces rust incidence under arid conditions. BioMed Research International. 2021;**2021**

[7] Hakim AR, Juraimi AS, Rezaul Karim SM, Khan MSI, Islam MS, Choudhury MK, et al. Effectiveness of herbicides to control rice weeds in diverse saline environments. Sustainability. 2021;**13**:2053

[8] Iqbal A, Iqbal MA, Awad MF, Nasir M, Sabagh A, Siddiqui MH. Spatial arrangements and seeding rates influence biomass productivity, nutritional value and economic viability of maize (*Zea mays* L.). Pakistan Journal of Botany. 2021;**53**(3):967-973

[9] Alam MA, Skalicky M, Kabir MR, Hossain MM, Hakim MA, Mandal MSN, et al. Phenotypic and molecular assessment of wheat genotypes tolerant to leaf blight, rust and blast diseases. Phyton, International Journal of Experimental Botany. 2021;**90**(4): 1301-1320

[10] Sorour S, Amer MM, El Hag D, Hasan EA, Awad M, Kizilgeci F, et al. Organic amendments and nano-micronutrients restore soil physico-chemical properties and boost wheat yield under saline environment. Fresenius Environmental Bulletin. 2021;**30**(9):10941-10950

[11] Boval M, Dixon RM. The importance of grasslands for animal production and other functions: A review on management and methodological progress in the tropics. Animal. 2012;**5**:748-762

[12] Ramankutty N, Evan AT, Monfreda C, Foley JA. Farming the planet: 1. Geographic distribution of global agricultural lands in the year 2000. Global Biogeochemical Cycles. 2008;**22**:1003-1025

[13] Han JG, Zhang YG, Wang CJ, Bai WM, Wang YR, Han GD, et al. Rangeland degradation and restoration management in China. The Rangeland Journal. 2008;**30**:233-239

[14] Brunstad RJ, Gaasland I, Vårdal E. Multifunctionality of agriculture: An

inquiry into the complementarity between landscape preservation and food security. European Review of Agricultural Economy. 2005;**32**(4): 469-488

[15] Zhao Y, Wu J, He C, Ding G. Linking wind erosion to ecosystem services in drylands: A landscape ecological approach. Landscape Ecology. 2017;**32**(12):2399-2417

[16] Zhang X, Niu J, Buyantuev A, Zhang Q, Dong J, Kang S, et al. Understanding grassland degradation and restoration from the perspective of ecosystem services: A case study of the Xilin River Basin in Inner Mongolia, China. Sustainability. 2016;**8**(7):594-603

[17] Yan Y, Xu X, Xin X, Yang G, Wang X, Yan R, et al. Effect of vegetation coverage on aeolian dust accumulation in a semiarid steppe of northern China. Catena. 2011;7(3): 351-356

[18] Winfree R, Fox JW, Williams NM, Reilly JR, Cariveau DP. Abundance of common species, not species richness, drives delivery of a real-world ecosystem service. Ecology Letters. 2015;**18**(7):626-635

[19] Wehn S, Hovstad KA, Johansen L. The relationships between biodiversity and ecosystem services and the effects of grazing cessation in semi-natural grasslands. Web Ecology. 2018; **18**(1):55-65

Interaction of Grassland Ecosystem with Livelihood and Wildlife Sustainability: Tanzanian Perspectives

Pius Yoram Kavana, John Kija Bukombe,

Hamza Kija, Stephen Nindi, Ally Nkwabi,

Iddi Lipende, Simula Maijo, Baraka Naftali,

Victor M. Kakengi, Janemary Ntalwila, Sood Ndimuligo

and Robert Fyumagwa

Abstract

In Tanzania, pure grasslands cover is estimated to be 60,381 km^2, about 6.8% of the total land area, and is distributed in different parts. These grasslands are diverse in dominant grass species depending on rainfall, soil type, altitude, and management or grazing system. They support livestock and wildlife distributed in different eco-tomes and habitats of the country. The potential of grasslands for the livelihood of rural people is explicit from the fact that local people depend solely on natural production to satisfy their needs for animal products. Analysis of grazing lands indicated that livestock population, production of meat, and milk from grasslands increased. But the wildlife population, when considered in terms of livestock equivalent units (Large Herbivore Units) showed a declining trend. The contribution of grasslands to the total volume of meat produced in the country showed a declining state, while milk production showed a slight increase. This situation entails a need to evaluate the grasslands of Tanzania to ascertain their potential for supporting people's livestock, wildlife, and livelihood. This study concluded that more research is needed to establish the possibility of grasslands to keep large numbers of grazing herbivores for sustainable livestock and wildlife production.

Keywords: grass species, grazing, livestock, Serengeti, Ugalla ecosystem, wildlife

1. Introduction

Grasslands are areas where the vegetation is dominated by grasses (Graminae species) and other herbaceous (non-woody) plants, having shrub or tree canopy cover not exceeding 2% (**Figure 1**) [1, 2]. Grasslands provide feed resources for grazing animals that include livestock and wildlife [3]. In addition, grasslands provide essential ecosystem services that include water catchments, biodiversity reserves, and socio-cultural and recreational needs [4, 5]. Grasslands are found in every continent and comprise 26% of the world's total land area and 80% of agricultural land and represent a wide variety of ecosystems [6].

In Sub-Saharan Africa, Angola, Benin, Botswana, Burkina Faso, Central African Republic, Cote d'Ivoire, Ethiopia, Ghana, Guinea, Kenya, Madagascar, Mozambique, Nigeria, Senegal, Somalia, South Africa, Tanzania, Zambia, and Zimbabwe have more than 100,000 km^2 of grassland [6]. According to Sulla-Menashe and Friedl [7] Moderate Resolution Imaging Spectroradiometer (MODIS) 2019 Land cover product (MCD12Q1) and International Geosphere-Biosphere Programme (IGBP) vegetation cover classes, the grasslands dominated by herbaceous annuals (<2 m) in Tanzania cover 385,427 km^2 which are distributed in different parts of the country.

Climate conditions and human activities affect the productivity of grasslands in Tanzania in terms of Net Primary Production (NPP) [8]. The general pattern of NPP in Tanzania shows a decreasing trend in the northeast-southwest, while the most significant decrease in NPP mainly occurred in the northeast [8, 9]. On the other hand, it predicted that the mean NPP values in the western, eastern, and central parts would increase by 2050 [8, 9]. Therefore, it implies an increase in the population of grazing animals with a consequential impact on people's livelihood in these areas. Thus, the prediction poses a need to establish baseline information on the capacity of grasslands to support livestock and wildlife with consequent effects on people's livelihood.

It is certain that grasslands provide numerous services and are central to the livelihoods and economies of many people in the country. Therefore, it is imperative to understand the current situation to develop strategies for sustaining this important biome. Therefore, this study was conducted to depict the importance of grasslands in Tanzania, their sustainability challenges, and how to keep productive grasslands in Tanzania.

Figure 1.
Typical grassland of Serengeti ecosystem in Northern Tanzania.

2. Methodology

2.1 Study area

This study covers all regions of Tanzania mainland as shown in the map (**Figure 2**).

2.2 Data collection and analysis

A systematic review of the scientific literature to obtain information on the grasslands of Tanzania was conducted using guidelines outlined by researchers [10, 11]. The study was done using various search engines, including Google Scholar, to establish the body of knowledge concerning the subject. The process involved a pre-defined search protocol using filters for keywords to audit search relevance and applicability [10]. The authors used experience from research conducted in Serengeti and Ugalla ecosystems and the eastern Tanzania grasslands to supplement the information obtained from the literature. The R software version 3.5.0 was used for data visualization with the ggplot2 package and analysis for correlation of human, wildlife populations and human-wildlife conflicts.

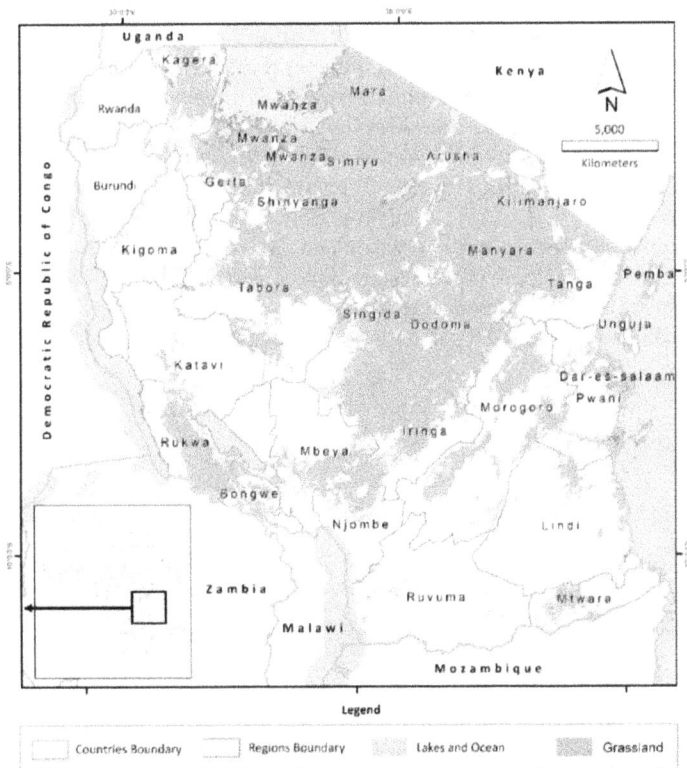

Figure 2.
Distribution of grasslands in Tanzania mainland (source: Tanzania Wildlife Research Institute GIS unit 2021).

3. Findings

Findings explained in this section combine ideas extracted from various sources of literature and authors' comprehensive understanding of the grasslands of Tanzania accrued by research experience.

3.1 Characteristics of Tanzania's grasslands

Grasslands are very diverse and widely distributed in Tanzania, with a range of dominant species depending on rainfall patterns, soil type, altitude, and management or grazing system. *Themeda triandra* is one of the most widespread grass species in Tanzania, and it is the dominant grassland type in central and northern Tanzania [9, 12]. However, the species is very variable and shows wide adaptation to growth in both the highlands and the lowlands. Themeda, Bothriochloa, Brachiaria, Sporobolus, Digitaria, and Heteropogon mixtures are common in the open dry areas such as the Serengeti plains (**Figure 3**).

Short tufted ecotypes of Themeda triandra are found at high altitudes and taller, more woody types are located in the open lowlands [9]. The dominant grass species in the drylands of central Tanzania include Cenchrus, Aristida, and Heteropogon. These grasses normally grow in association, and the establishment pattern of herbaceous plants is generally determined by the environment and soil texture [13]. Hyparrhenia, Hyperthelia, and *Pennisetum polystachyon* tall grass are common in the miombo woodlands of western Tanzania. The miombo forest is a vital vegetation type covering the southern two-thirds of Tanzania [14].

Pennisetum grasslands are classified into two types: high altitude grasslands of *Pennisetum clandestinum* and savannah grasslands of *Pennisetum mezianum* and *Pennisetum purpureum* [9, 15]. *P. clandestinum* is a prostrate stoloniferous perennial grass that is widely distributed in high altitudes (1400 m to over 3000 m.a.s.l *P. purpureum* is a tall, erect, vigorous perennial species that grows in damp grasslands and forest areas up to 2400 m.a.s.l.). At the same time, *Pennisetum mezianum* occurs in soils with impeded drainage heavy clay soil (Kavana, personal observation). *Panicum maximum* is a common grass in the eastern part of Tanzania and it is often associated with Bothriochloa in abandoned sisal farms in the coastal areas (Kavana, personal observation). The Panicum-Hyparrhenia is recognized as a region along the Coast

Figure 3.
Mixed plant species in open wooded grassland of western Serengeti, Tanzania.

northwards from Tanzania, Kenya, and finally into Somalia [15]. *P. maximum* is more typical grass of shady places in the foothills of Mountain ranges up to 2000 m.a.s.l. and is a pioneer grass that comes in after clearing and cultivating the lowland forest [15]. The Sporobolus-dominated grasslands usually exist on seasonally dry alkaline soils and are not destroyed by fire [16]. Therefore, grassland habitats provide valuable pastures in semi-arid areas of Tanzania where *Sporobolus pyramidalis*, *S. marginatus*, *S. ioclados* and *S. cordofanus* sometimes occur in association. *S. consimilis* and *S. spicatus* association occurs as a mosaic along the lakeshore for example along Lake Burunge in northern Tanzania [16].

In southern miombo woodlands of Tanzania, *T. triandra* is a dominant grass and widespread and occurs at different topographic positions [17]. On deeper Plateau soils tall grasses of *Hyperthelia dissoluta* and *Andropogon gayanus* of about 2 m dominate. On hill slopes, *Hyparrhenia newtonii* and *Andropogon schirensis* with a height of about 1.2–1.4 m is very frequently present. On leached soils grasses are mostly 0.6–0.8 m, *Aristida adscensionis* is primarily dominant [17].

In the western miombo woodlands of Tanzania, there are extensive, continuous woodlands interspersed with seasonally inundated grasslands known as "mbuga" in the Kiswahili language (**Figure 4**).

According to an unpublished report by Kavana and Kakengi [18] they observed that common grasses in seasonally flooded plains include *Cynodon dactylon, C. articularis, C. cyperoides, C. difformis, C. dives, C. esculantus, C. involucratus, Cyperus papyrus, C. rotundus, Phragmates mauritianus, Pennisetum purpureum, Sporobolus spp, Echinochloa pyramidalis* and *Oryza longistaminata* (**Figure 5**).

In settlement areas where miombo woodlands are cleared for agricultural and grazing lands, the common grasses include *Panicum maximum, P. repens, Panicum trichocladum, P. trichocladum, Pennisetum polystachyon, P. polystachyon, Rhynchelytrum repens, Setaria homonyma, S. sphacelata, Sporobolus africanus, S. fimbriatus, S. ioclados, Sporobolus pyramidalis, S. sanguineus, Urochloa decumbents* and *U. echinolaenoides* (**Figure 6**).

Some characteristic mosaic grassland occurs proximal to Lake Rukwa where dominant grass species stand a change with soil alkalinity progressing towards the lake [19]. Cymbopogon begins in less alkaline soil (pH 7.5–8.5) followed by Hyparrhenia, Chloris, Cynodon, *Sporobolus robustus*, Echinochloa, Cyperus, Diplachne, and *Sporobolus spicatus* at pH 9.5–11 [19]. Therefore, Tanzania is diverse and influenced by climatic conditions, soil, and anthropogenic activities [8].

Figure 4.
Typical characteristics of grassland in Ugalla ecosystem assessed by researchers from Tanzania Wildlife Research Institute.

Figure 5.
Seasonally flooded grassland plains of Western Serengeti were assessed by researchers from Tanzania Wildlife Research Institution.

Figure 6.
Robust Pennisetum polystachyon and Rhynchelytrum repens in agricultural land, western Tanzania.

3.2 Importance of grasslands for grazing animals

Grasslands can be classified as natural and improved grasslands. Natural grasslands are dominated by native grass species mixtures that occur naturally, while improved grasslands are developed by seeding and vegetative propagation of selected grass species [20]. Grasses form a basal diet for both livestock and wildlife animals that make livestock production in the traditional sector and the most protected areas for wildlife to rely on [21]. This situation leads to Livestock-wildlife competition that operates through two sets of processes within the social-ecological systems and economic processes that influence Livestock and wildlife-based enterprises as sources of income for people and the nation, respectively [20]. The authors considered that ecological processes affect the relative efficiencies of livestock and wildlife species in utilizing grasslands' feed and water resources. As a result, the contribution of livestock enterprises to people's livelihood generally exceeds the contribution of wildlife conservation to the livelihood of people within the same area [22]. Further studies show that economic processes result typically in agricultural returns to outcompete wildlife returns and the patchwork of land use within rangelands intensifies towards croplands and fragmented rangelands [22, 23]. This scenario corroborates observations made in the grasslands of Tanzania, as shown in **Figure** 7.

Population trends in **Figure** 7 indicate that livestock increased continuously from 1995, possibly due to an increase in demand for livestock products that resulted from the increase in the human population. It has been reported that there is a close relationship between increases in cattle and growth in the human population that result from the need for livestock products to cater to the growing human population [26]. However, livestock in grasslands increases typically at the expense of wildlife [26]. Many studies show that markets, technology, and infrastructure development, the position of a rangeland/grassland on its production possibility frontier (PPF), changes with agricultural production becoming specialized, driving down the possibilities for wildlife production [27–31]. The overall population trend of grazers in the grasslands of Tanzania indicates a steady increase in the population [32]. This implies an increase in grazing pressure in grasslands that entails the need for close monitoring of the grasslands of Tanzania for sustainable livestock and wildlife production [8].

Frequent and severe droughts in many parts of Tanzania are being felt with their associated consequences on food production and water scarcity, leading to food shortages and insecurity, water scarcity, hunger, and it provides poor forage for animals [33]. Prolonged drought is a significant driver of grassland ecosystems and is likely to lead to increased wildfires and loss of wetland habitats that are critical habitats for migratory bird species and species migration and habitat shifts [34]. An increase in temperatures, reduced rainfall, and drought is already being observed in some regions of Tanzania [35]. In particular, the northern part of Tanzania-central Serengeti grassland is projected to become even drier in this century [36]. In wetter areas, forests are likely to encroach on existing grasslands. In contrast, deserts are projected to expand in extent and move upward in elevation in increasingly arid areas, causing "desertification" of arid grassland ecosystems. Such a process will greatly affect the productivity of the grasslands ecosystem, impacting animal welfare [35, 36].

3.2.1 Natural grasslands

Survival of wildlife in protected areas and livestock production in the traditional sector in Tanzania rely on grasslands dominated by native grass species. The native grass species provide a basal diet for both wildlife and livestock herbivores. Native

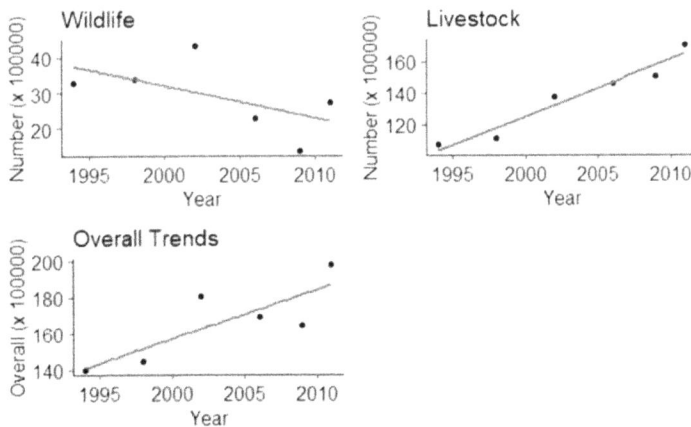

Figure 7.
Large Herbivore Units (LHU) population for livestock and wildlife in grasslands of Tanzania (source: Authors' computation based on TZFAOSTAT_data [24, 25]).

grass species inherently vary in biomass and nutrient contents they supply to the grazing animals. This compels wildlife and livestock to select certain grass species when grazing to meet their energy and nutrient requirements [37]. For a grass species to be consumed by grazing animals, it must belong to edible plant species (i.e., not harmful). Among edible plant species, some species are highly desirable, desirable, and less desirable. Plant species that are not edible are termed undesirable plants in terms of grazing animals' feeding value [37]. The natural grasslands of Tanzania are mainly composed of desirable and highly desirable grass species (**Figure 8**). This composition supports the survival of large numbers of grazing wildlife and livestock in the country.

Nutritive value of grasses in natural grasslands vitiates in quality rapidly across months within a year (**Figure 9**). This makes it rather difficult for natural grassland to support the high productivity of grazing animals throughout the year. High grazing animal production and products follow periods of high quantity and quality of grasses in grasslands [39]. In other words, there is natural synchronization of the reproduction cycle such that most of the calving occurs during periods of high quantity and quality of grasses in natural grasslands [39].

3.2.2 Improved grasslands

Improved grasslands in Tanzania are classified according to usages, such as pasture production farms for haymaking and grazing farms for dairy production. These farms are seeded with improved grass species: *Chloris gayana*, *Cenchrus ciliaris*, *Pennisetum purpureum*, *Panicum maximum*, *Setaria sphacelata*, *Tripsacum laxum* and improved varieties of Brachiaria species. In some cases, leguminous species are over-sowed in grass farms to improve the quality of hay. Common leguminous species mixed with grass include *Desmodium uncinatum*, *Centrosema pubescens*, *Macroptilium atropurpureum*, *Stylosanthes guianensis*, *S. guianensis*, *Pueraria phaseoloides*, *Clitoria ternatea* and *Calopogonium mucunoides*.

A considerable amount of work for improved grassland was carried out in Tanzania at Kongwa Pasture Research Centre with large-scale sowing of *C. ciliaris* under large-scale management [40–42]. Material initially selected in Tanzania was much more widespread use outside the country: common cultivars of tropical

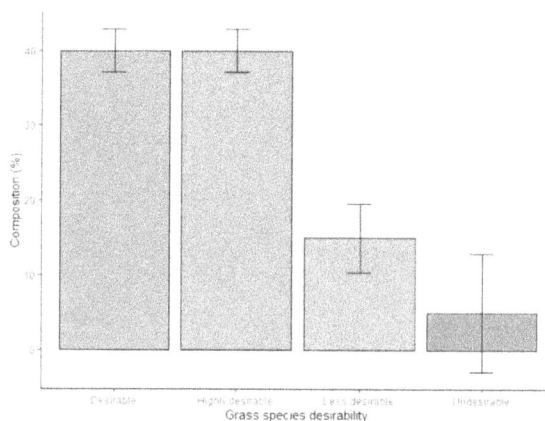

Figure 8
The desirability of grass species in western Serengeti natural grassland. Source: Authors' computation based on Kavana et al. [38].

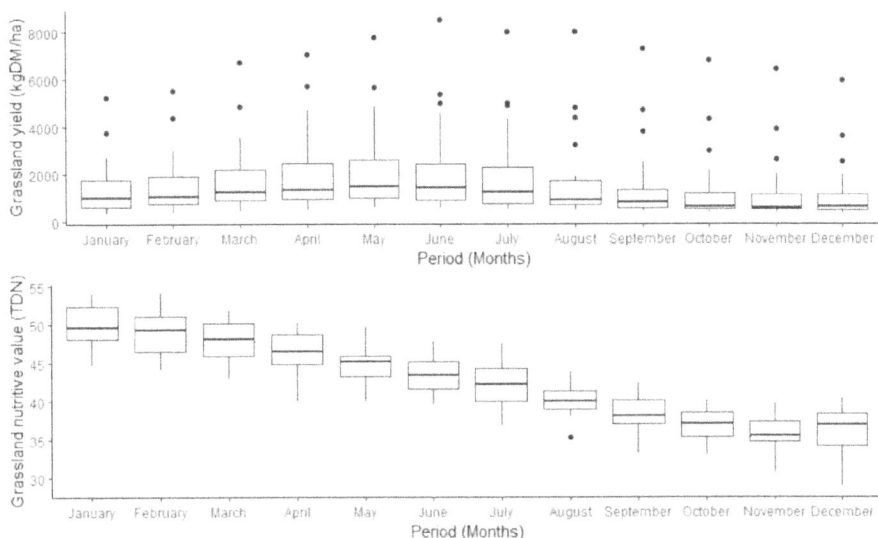

Figure 9.
Quantity and quality of grasses in Ugalla ecosystem's natural grasslands. Source: Authors computation from unpublished research data.

pasture plants developed from Tanzanian material in Australia include *C. ciliaris* "Biloela", *C. gayana* 'Callide' and *Neonotonia wightii* 'Clarence' [43]. Mixing grasses with legumes is considered to increase dry matter production of grass. The highest dry matter yield of *C. ciliaris* was observed in a mixture with *Phaseolus atropurpureus*, *S. guianensis* and *C. pubescens* [44].

3.3 Contribution of grasslands to the livelihood of rural communities

A direct result of the contribution of grassland to the livelihood of people and the national economy in Tanzania is that local people rely mainly on grassland for the production of livestock products. The trend of meat production from grasslands (**Figure 10**) shows that meat production increased at a decreasing rate, and the value of meat produced from grasslands increased progressively. This indicates that livestock keeping in grasslands is a lucrative business that contributes to rural people's economy. However, the contribution of meat produced from grassland to the total meat produced by grazing livestock showed a declining trend. This could be attributed to the decline in the potential of Tanzania's grassland of Tanzania to support large herds of grazing animals. The potential of grasslands to support grazing animals is affected by environmental fluctuations and increased human activities [45, 46]. Human population increase resulted in the expansion of cultivated land at the expense of grassland, and the need for animal products led to the keeping of large herds of livestock [26]. This situation causes shrinkage of grassland and overgrazing, reducing grassland's potential to support grazing animals in the country.

Grasslands contribute more than 60% of milk produced in Tanzania, and the value of milk produced increased steadily (**Figure 11**), contributing more than 2500 billion TZS to people's economy. However, the contribution of grassland to the total milk produced in Tanzania has been slightly increasing year after year [47]. This might be caused by improvement in urban dairy farming and probably deterioration of grasslands in terms of quantity and quality of feed resources availability in communal grasslands that need to be evaluated.

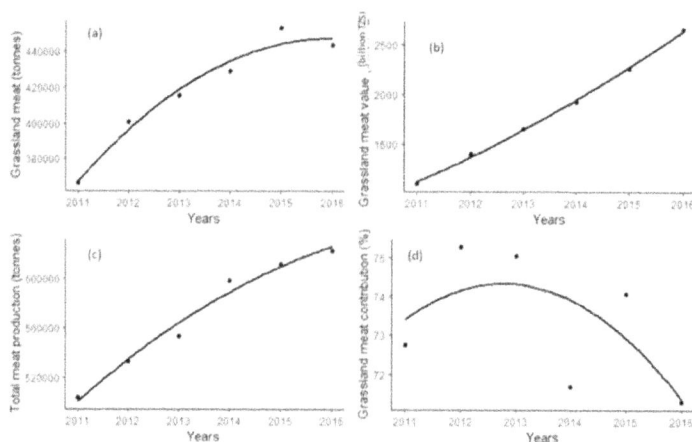

Figure 10.
*Quantity and value of meat produced from grasslands and other places of Tanzania. Source: Authors'
computation based on livestock and fisheries basic data and Tanzania in figures 2016 documents.*

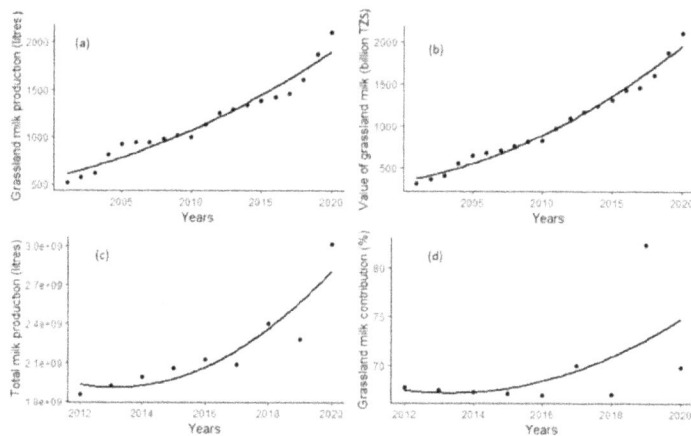

Figure 11.
*Contribution of grassland to milk production and economy of people in Tanzania. Source: Authors'
computation based on Tanzania in figures document [47].*

3.4 Challenges on the sustainability of grasslands

Agriculture poses a significant challenge for the sustainability of grasslands in
Tanzania. Current agriculture production is hinged on the expansion of mono-
cropping farms to increase food production for the growing human population. In
many cases, the expansion of crop farms is done at the expense of grasslands. Land
clearing and cultivation for crop production re-structure and disrupts a previously stabilized grassland ecosystem. The disturbed ecosystem due to cultivation
immediately begins succession where annual grasses and forbs adapted to bare land
conditions and disturbed soil invade the site and become established. This situation
results in changes in plant species composition of the grassland.

Grasslands are the main grazing areas for livestock where most of the grazing
lands in the country are communally managed. Poor livestock grazing management
results in soil compaction due to the effects of animal trampling, leading to poor
water infiltration. Removal of plants due to many grazing livestock in communal

lands causes bare land (**Figure 12**). In combination with poor water, infiltration causes surface water runoff during the rainy season that erodes soil.

Soil erosion of bare land during the rainy season removes top fertile soil resulting in low soil fertility. Poor soil fertility causes the establishment of a limited number of plant species resulting in low plant composition on grassland. Low plant species composition leads to low above-ground biomass production that causes an insufficient feed resource base for grazing animals.

A high number of grazing animals in a shrinking grassland with an insufficient feed resource base in terms of quantity and quality results in high utilization pressure by grazing animals. High grazing pressure exerted on highly desirable and desirable grass decreases the potential of grassland to support grazing animals.

Effects of climate change on the grasslands of Tanzania could be manifested in relation to the variability of temperature and precipitation [8]. Climate change projections indicate that western parts of the country, central, north, Lake Victoria basin, eastern parts of Lake Nyasa, south-western and north-eastern highlands are projected to feature increased minimum temperature [48]. An increase in temperature is associated with an increase in the photosynthetic rate of plants under optimum soil moisture. Projections of precipitation indicate that Coastal regions, parts of north-eastern highlands, northern areas, western and southern parts of the Lake Victoria basin will experience an increase in annual rainfall. This implies that most of the grasslands of Tanzania are expected to increase Net Primary Production (NPP) of grass biomass. Therefore, the main limitations for the sustainability of most grasslands of Tanzania will be extensive crop cultivation and livestock population grazing.

3.5 Way forward to sustain grasslands

An increase in the human population triggers a need for food sufficiency that stimulates cultivation of land and livestock keeping which is based on the availability of land. This shows that in many cases increase in the human population in rural areas results in the expansion of cultivated land to cater to food demand. However, this scenario occurs when land is available for expansion of crop farms but changes to intensive cultivation when land is scarce.

Figure 12.
Bare land due to overgrazing and trampling in the grassland of western Serengeti.

Changing from extensive crop cultivation to intensive cultivation and reducing grazing pressure is inevitable for the sustainability of grasslands in Tanzania because the land is a fixed commodity. Intensive cultivation will be achieved by increasing production in the same land units by increasing agricultural inputs and technology. Reducing grazing pressure in grasslands will be achieved by either reducing the number of grazing animals per unit of land or reducing the duration of grazing on grassland units.

Determinants of the holistic direction that agro-pastoralism is developing in Tanzania are still not clearly understood. A combined model including economic, social, ecological components and wildlife conservation is needed to enable predictions about the future of agro-pastoralism in areas that are adjacent to protected areas.

4. Conclusions

Our review details the significance of grasslands in Tanzania. Grasslands of Tanzania are diverse, and their diversity is influenced by both climatic, soils, and topographic variations across Tanzania. They support an enormously larger number of livestock and wildlife populations.

Grasslands of Tanzania show a big potential to support people's livelihood through meat and milk production, but its contribution is declining. However, such ecosystems experience overgrazing, conversion to agricultural lands, frequent annual fires, and climate change.

Evaluation of the grasslands of Tanzania is worthwhile to establish baseline information or trends that will be used for comparison in long-term monitoring of grasslands condition. This can only be achieved if proper land use plans are put in place, especially in all rural settings and improved management systems that operate at different regimes across Tanzania.

Evaluation of the grasslands of Tanzania is inevitable to establish baseline information for comparison in long-term monitoring of grasslands condition. However, we can only achieve this if proper land use plans are implemented, especially in rural settings.

Conflict of interest

The authors proclaim no conflict of interest.

Author details

Pius Yoram Kavana[1*], John Kija Bukombe[2], Hamza Kija[2], Stephen Nindi[2],
Ally Nkwabi[1], Iddi Lipende[1], Simula Maijo[1], Baraka Naftali[1], Victor M. Kakengi[2],
Janemary Ntalwila[2], Sood Ndimuligo[3] and Robert Fyumagwa[2]

1 Tanzania Wildlife Research Institute, Mahale-Gombe Wildlife Research Center,
Kigoma, Tanzania

2 Tanzania Wildlife Research Institute, Head Quarters, Arusha, Tanzania

3 Center for Ecological and Evolutionary Synthesis (CEES), Department of
Biosciences, University of Oslo, Norway

*Address all correspondence to: pius.kavana@tawiri.or.tz

IntechOpen

References

[1] Peeters A, Beaufoy G, Canals RM, De Vliegher A, Huyghe C, Isselstein J, et al. Grassland term definitions and classifications adapted to the diversity of European grassland-based systems. Grassland Science in Europe. 2014;**19**:743-750

[2] Allen VG, Batello C, Berretta E, Hodgson J, Kothmann M, Li X, et al. An international terminology for grazing lands and grazing animals. Grass and Forage Science. 2011;**66**(1):2-29. DOI: 10.1111/j.1365-2494.2010.00780.x

[3] Venter ZS, Hawkins H-J, Cramer MD. Cattle don't care: Animal behaviour is similar regardless of grazing management in grasslands. Agriculture, Ecosystems & Environment. 2019;**272**:175-187

[4] Turpie J, Forsythe K, Knowles A, Blignaut J, Letley G. Mapping and valuation of South Africa's ecosystem services: A local perspective. Ecosystem Services. 2017;**27**:179-192. DOI: 10.1016/j.ecoser.2017.07.008

[5] Baskent EZ. A framework for characterizing and regulating ecosystem services in a management planning context. Forests. 2020;**11**(1):102. DOI: 10.3390/f11010102

[6] White RP, Murray S, Rohweder M. Pilot Analysis of Global Eosystems: Grassland Ecosystems. Washington, D. C.: World Resources Institute; 2000. p. 81

[7] Sulla-Menashe D, Friedl MA. User Guide to Collection 6 MODIS Land Cover (MCD12Q1 and MCD12C1) Product. Reston, VA, USA: USGS; 2018;**39**:1-18

[8] Zarei A, Chemura A, Gleixner S, Hoff H. Evaluating the grassland NPP dynamics in response to climate change in Tanzania. Ecological Indicators. 2021;**125**:107600. DOI: 10.1016/j.ecolind.2021.107600

[9] Rattray JM. The Grass Cover of Africa. Rome, Italy: FAO; 1960

[10] Pullin AS, Stewart GB. Guidelines for systematic review in conservation and environmental management. Conservation Biology. 2006;**20**:1647-1656

[11] Inskip C, Zimmermann A. Human-felid conflict: A review of patterns and priorities worldwide. Oryx. 2009; **2009**(43):18-34

[12] Snyman HA, Ingram LJ, Kirkman KP. Themeda triandra: A keystone grass species. African Journal of Range & Forage Science. 2013;**30**(3):99-125

[13] Kavana PY, Mtengeti EJ, Sangeda A, Mahonge C, Fyumagwa R, John B. How does agro-pastoralism affect forage and soil properties in western Serengeti, Tanzania? Tropical Grasslands-Forrajes Tropicales. 2021;**9**(1):120-133. DOI: 10.17138/TGFT(9)120-133

[14] Jew EKK. Rapid land use change, biodiversity and ecosystem services in miombo woodland: Assessing the challenges for land management in south-west Tanzania [doctoral dissertation]. West Yorkshire, England: University of Leeds; 2016

[15] Herlocker D, editor. Rangeland Ecology and Resource Development in Eastern Africa. Nairobi, Kenya: GTZ; 1999

[16] Vesey-Fitzgerald. The origin and distribution of valley grasslands in east Africa. Journal of Ecology. 1970;**58**(1):51-75

[17] Bloesch U, Mbago F. Vegetation Study of Selous-Niassa Wildlife Corridor for biodiversity, conservation values and management strategies. In: Consultancy Report Submitted to Wildlife Division. Ministry of Natural Resources and Tourism, United Republic of Tanzania; 2006. p. 70

[18] Kavana PY, Kakengi VAM. Annual report. Feed abundance and diversity for livestock in communities surrounding the Ugalla ecosystem. Unpublished Report Submitted to Tanzania Commission for Science and Technology (COSTECH). Dar-es-Salaam, Tanzania; 2009. p. 10

[19] Dean GJW. Grasslands of the Rukwa Valley. Journal of Applied Ecology. 1967;**4**:45-57

[20] du Toit JT, Cross PC, Valeix M. Managing the livestock–wildlife interface on Rangelands. In: Briske D, editor. Rangeland Systems. Cham: Springer Series on Environmental Management. Springer; 2017

[21] Milchunas DG, Sala OE, Lauenroth WK. A generalized model of the effects of grazing by large herbivores on grassland community structure. The American Naturalist. 1988;**132**(1):87-106

[22] Ogutu JO, Piepho HP, Said MY, Ojwang GO, Njino LW, et al. Extreme wildlife declines and concurrent increase in livestock numbers in Kenya: What are the causes? PLoS One. 2016;**11**(9):e0163249. DOI: 10.1371/journal.pone.0163249

[23] Hobbs NT, Galvin KA, Stokes CJ. Fragmentation of rangelands: Implications for humans, animals, and landscapes. Global Environmental Change. 2008;**18**:776-785

[24] United Republic of Tanzania (URL). Ministry of Natural Resources and Tourism. Wildlife Sub-Sector Statistical Bulletin. 2 ed. Dar-es-Salaam, Tanzania: National Bureau of Statistics; 2013. p. 77

[25] Peden DG. Livestock and wildlife population distributions in relation to aridity and human populations in Kenya. Journal of Range Management. 1987;**40**(1):67-71

[26] Kavana PY, Mahonge CP, Sangeda AZ, Mtengeti EJ, Fyumagwa R,

Nindi S, et al. Panorama of agro-pastoralism in Western Serengeti: A review and synthesis. Livestock Research for Rural Development. 2017;**29**:Article #191. Available from: http://www.lrrd.org/lrrd29/10/pyka29191.html

[27] Bastian CT, Jacobs JJ, Held LJ, Smith MA. Multiple use of public rangeland: Antelope and stocker cattle in Wyoming. Journal of Range Management. 1991;**44**:390-394

[28] Smith PF, Gorddard R, House APN, McIntyre S, Prober SM. Biodiversity and agriculture: Production frontiers as a framework for exploring trade-offs and evaluating policy. Environmental Science & Policy. 2012;**23**:85-94

[29] Altchenko Y, Villholth KG. Trans boundary aquifer mapping and management in Africa: A harmonized approach. Hydrogeology Journal. 2013;**21**:1497-1517

[30] Copeland HE, Pocewicz A, Naugle DE, Griffiths T, Keinath D, et al. Measuring the effectiveness of conservation: A novel framework to quantify the benefits of sage-grouse conservation policy and easements in Wyoming. PLoS One. 2013;**8**(6):e67261. DOI: 10.1371/journal.pone.0067261

[31] Northrup JM, Wittemyer G. Characterizing the impacts of emerging energy development on wildlife, with an eye towards mitigation. Ecology Letters. 2013;**16**:112-125

[32] Okello MM, Kenana L, Maliti H, Kiringe JW, Kanga E, Warinwa F, et al. Population status and trend of water dependent grazers (buffalo and waterbuck) in the Kenya-Tanzania Borderland. Natural Resources. 2015;**6**(02):91-114. DOI: 10.4236/nr.2015.62009

[33] Mariotti E, Parrini F, Louw CJ, Marshal JP. What grass characteristics drive large herbivore feeding patch selection? A case study from a South

African grassland protected area. African Journal of Range & Forage Science. 2020;**37**(4):286-294

[34] Holdo RM, Fryxell JM, Sinclair AR, Dobson A, Holt RD. Predicted impact of barriers to migration on the Serengeti wildebeest population. PLoS One. 2011;**6**(1):e16370. DOI: 10.1371/journal.pone.0016370

[35] Nkwabi A, Mgimwa E, John J, Ombeni E, Kamugisha E. International Waterbird Census of Tanzania: National Report, Unpublished Report, March 2021. Tanzania Wildlife Research Institute; 2021

[36] Nkwabi AK. Influence of seasonal habitat variation and agriculture on abundance, diversity and breeding of birds in Serengeti National Park and surrounding areas [doctoral dissertation]. Tanzania: University of Dar es Salaam; 2017

[37] Bukombe J, Kittle A, Senzota RB, Kija H, Mduma S, Fryxell JM, et al. The influence of food availability, quality and body size on patch selection of coexisting grazer ungulates in western Serengeti National Park. Wildlife Research. 2019;**46**(1):54-63

[38] Kavana PY, Sangeda AZ, Mtengeti EJ, Mahonge C, Bukombe J, Fyumagwa R, et al. Herbaceous plant species diversity in communal agro-pastoral and conservation areas in Western Serengeti, Tanzania. Tropical Grasslands-Forrajes Tropicales. 2019;**7**(5):502-518. DOI: 10.17138/TGFT(7)502-518

[39] Mcnaughton S. Grazing as an optimization process: Grass-ungulate relationships in the Serengeti. The American Naturalist. 1979;**113**(5):691-703

[40] Sarwatt S, Mollel E. Country Pasture/Forage Resource Profiles—United Republic of Tanzania. Rome, Italy: FAO; 2006. Available from: http://www.fao.org/ag/aGp/agpc/doc/Counprof/PDF%20files/Tanzania_English.pdf [Accessed: September 28, 2021]

[41] Brzotowski HW. Influence of pH and superphosphate on establishment of Cenchrus ciliaris from seed. Tropical Agriculture Trinidad and Tobago. 1962;**39**:289-296

[42] Brzotowski HW, Owen MA. Botanical changes in sown pasture. Tropical Agriculture Trinidad and Tobago. 1964;**41**:231-242

[43] Owen MA, Brzostowski HW. A grass cover for the Upland soils of the Kongwa plain, Tanganyika. Tropical Agriculture Trinidad and Tobago. 1966;**43**:303-314

[44] Walker B. Effects of nitrogen fertilisers and forage legumes on Cenchrus ciliaris pasture at Ukiriguru. East African Agricultural and Forestry Journal. 1967;**35**:2-5

[45] Fang Y, Xiangzheng D, Qin J, Yongwei Y, Chunhong Z. The impacts of climate change and human activities on grassland productivity in Qinghai Province, China. Frontiers in Earth Science. 2014;**8**(1):93-103. DOI: 10.1007/s11707-013-0390-y

[46] Seddon AW, Macias-Fauria M, Long PR, Benz D, Willis KJ. Sensitivity of global terrestrial ecosystems to climate variability. Nature. 2016;**531**:229. DOI: 10.1038/nature16986

[47] National Bureau of Statistics (NBS). The United Republic of Tanzania 2020. Tanzania in Figures. 2020. Available from: National Bureau of Statistics - Tanzania in Figures 2020 (nbs.go.tz)

[48] Luhunga PM, Kijazi AL, Chang'a L, Kondowe A, Ng'ongolo H, Mtongori H. Climate change projections for Tanzania based on high-resolution regional climate models from the coordinated regional climate downscaling experiment (CORDEX)-Africa. Frontiers in Environmental Science. 2018;**6**:122. DOI: 10.3389/fenvs.2018.00122

Chapter 3

Earth's Energy Budget Impact on Grassland Diseases

Ang Jia Wei Germaine

Abstract

The change in climate have caused different biotic and abiotic factors to be more prominent when management plan is executed. The increase in temperature have then cause frequent drought that may attract alien species of vectors to spread novel diseases among the native plants. However, the change in climate varies in different countries. Thus, common diseases that threatens food security such as *Xanthomonas spp.*, *Pseudomonas spp* are in limelight of research. Vectors lifecycle may cause plant diseases to by cyclative. Therefore, to find the break in the vector's lifecycle will be a method to eradicate harmful population in grassland. Modern days will then call for innovative method and limitations should be considered. Climate change have also impacted pathogens migration and mating pattern. The need for innovative management is constantly on the rise.

Keywords: Climate Change, insects, fungus, viruses, grassland, diseases, bacteria, vectors

1. Introduction

The change in climate have impacted the grassland in many ways. Grassland have the capability to buffer climate variability. They provide many other services to the ecosystem as well. The change in earth's energy budget calls for innovative methods to manage the loss of grassland. Understanding the importance of the presence of grassland, the need to manage loss and be economically efficient is crucial as well.

Atmospheric warming and climate change have the potential for significant effects on agriculture systems and their productivity. Crops and forage systems have display significant vulnerability as the change in temperature and precipitation will then impact cultivation, sowing, growth and utilisation [1]. Farmers will then have to innovate and have other management methods to counter the effects of climate change.

Climate variability have caused frequent droughts. This have impacted the grassland by increasing plant mortality and limiting the geographic distribution of plant species, accelerating grassland degradation [2]. In addition to the observable change, there are other biotic and abiotic factors that will be affected as well. The microorganisms that live in the soil biota changes. The biodiversity may decrease and alien species may increase. With the change in biodiversity in grassland, novel diseases in plant may arise. Grassland diseases are a major part of grassland management. To understand the underlying physiology of pathogens and mode of transmission will be crucial, as intercepting at the point of weakness of pathogen's lifecycle can reduce damages to vegetation and other costs for management [3].

Different type of grasslands across the globe will have different management requirements as the difference in pathogens differ as the environmental factors differs. The imbalance of Earth's energy budget will further complicate the understanding and requirements for grassland management. Therefore, this chapter aims to cover and understand how did climate change impact the components that cause plant pathogens to continue to cause damage to grassland. In addition, the chapter covers the common types of grassland diseases that have been a recurring problem in various grasslands and its causative agents.

2. Earth's energy budget

Energy cannot be created nor destroyed. Earth will require solar energy in order for the basis of life to continue. This can be evidently observed by plants requiring sunlight for photosynthesis to occur and to produce oxygen for living organisms. Earth would freeze without sunlight. The ideal balance of Earth' energy budget can be explained with the guidance of the diagram below.

In summary, the incoming solar energy is being used, reflected and radiated back to space. To achieve the ideal earth's energy budget, the incoming solar energy will be equal to the outgoing solar energy (which includes energy that have been reflected back into space).

The earth's energy is constantly changing as the energy flows through the system. The changes in earth' energy balance have been contributed by the components human activities. This causes changes to the composition of the atmospheric layers. As such, this could lead to the increased absorption of radiation or decreased absorption of radiation by reflecting those energy back into space as there is high albedo in the atmospheric layer. Albedo is an elaborate word that has a simple physical concept. Lighter surfaces on earth reflects more heat than dark surfaces. Earth's energy budget in the past was balanced by the long wavelength that is being absorbed and the short wavelength that is being reflected back into the solar system. The reflection of short waves energy could be emitted by earth's surfaces, clouds, atmosphere, conduction and/or convections and, evapotranspiration. With the imbalance of absorption and reflection it could cause a positive energy imbalance, Earth system is said to be gaining energy causing global warming. With the continuous gain in energy, the albedo in earth would decrease as the ice caps and snow starts melting.

Global warming increases not only the global temperature. The concentration levels of greenhouse gases and of those gases increase, carbon dioxide is of interest to a lot of scientist. The increase in carbon dioxide have been contributed by human activities such as deforestation and burning of fossil fuels (just some to name). The five carbon pools that will cycle the concentration of carbon in the Earth were lithosphere, oceans, soil organic matter, atmosphere and biosphere. Oceans are the biggest carbon pool in the Earth. However, deforestation has contributed to the global temperature rise as deforestation will cause a decreased in absorption of carbon dioxide. With the increase in carbon dioxide in the atmosphere have caused sun's radiation is being reflected back to earth rather than back into space. Hence, as the Earth loss the ability to release energy, the global temperature increase. Apart from carbon dioxide, the increase in concentration for other atmospheric gases will allow different wavelength of light to pass through.

Thus, greater the amount of atmospheric gases that absorb thermal infrared radiation from the Earth's surface, the greater the proportion of radiation emitted from the atmosphere towards the Earth's surface [4]. This would then result in the Earth's surface being less negative. More energy is then available for sensible and latent heat flux at the surface. Thus, the increase in air temperature.

The change in earth's energy budget does not impact solely on the plants in Grassland. It would also impact those that are living in the grassland. The impact of global warming stresses the ecosystems through several changes that could already be experienced: rise in global temperature, water shortages, drought and intense storm damage. In addition to those that have been experiences, salt invasion is a rising problem. The influx of salt into the soil and water can change the ionic concentration of an area. The sudden change in soil environment will give little time for underground organisms to adapt.

3. Methodology

The measurement of the Earth's energy budget can be conducted through remote sensing. A review by Liang et al. [5] has mentioned that there are several components to be calculated. The first formula was to get the surface energy balance and this is the sum of soil heat flux (G), sensible heat flux and latent heat flux. The latent heat flux is derived from the product of latent heat evaporation of water and the rate of evaporation of water.

$$R_n = G + H + \lambda ET \tag{1}$$

R_n is the representation of all-wave net radiation.

However, remote sensing has presented another perspective, where the net radiation is the sum of shortwave net radiation and long wave net radiation (which is represented in **Figure 1**).

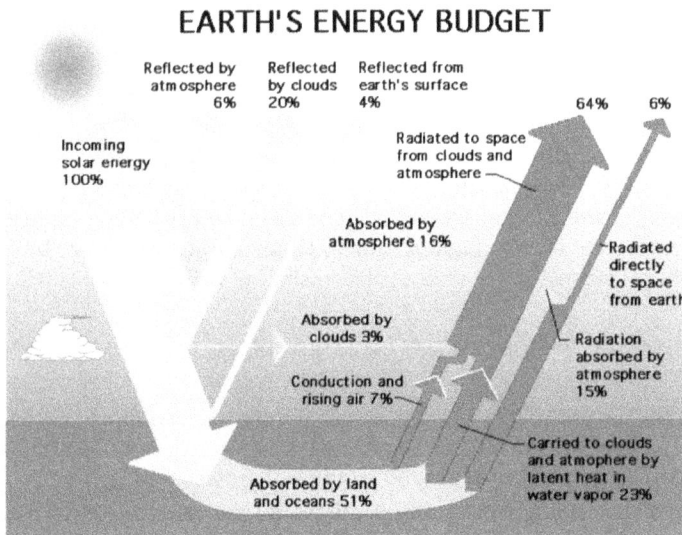

Figure 1.
The diagram depicts an overall movement of solar energy where the energy dissipates into space, being retained on the surface of earth and those reflected by the atmosphere. Courtesy of the NASA global precipitation measurement education.

$$R_n = R_n^s + R_n^l + \left(1-\alpha_{sw}\right)F_d^s + F_d^l - F_u^l = \left(1-\alpha_{sw}\right)F_d^s + \mathcal{E}F_d^l - \sigma\varepsilon T_s^4 \qquad (2)$$

The equation above will then include all the other factor that will affect the energy balance. The net radiation is simply the sum of shortwave net radiation (R_n^s) and long wave net radiation (R_n^l). However, the incoming waves will then be affected by the albedo on earth.

Hence, the second part of the equation WHERE the product of the difference in surface shortwave broadband albedo $\left(1-\alpha_{sw}\right)$ and the shortwave downward flux incident on the surface (F_d^s), in addition to the difference between the longwave downward (F_d^l) and upwelling radiation (F_u^l) will give the all-wave net radiation.

The third part of the equation will then include the Stefan-Boltzmann's constant (σ). The product of $\left(1-\alpha_{sw}\right)F_d^s$ is then added to the product of surface longwave broadband emissivity ($\mathcal{E}$). The product of and the skin surface temperature $\left(T_s^4\right)$ will be deducted and this gives the all-wave net radiation.

The remote sensors on the satellite have been used to measure the Total Solar Irradiance. The sensors from previous studies have allowed scientists to estimate solar constant. Remote sensing on the satellite has the ability to sense the net radiation at the top of the atmosphere. The data recorded includes both spatial and temporal scales. Remote sensing have been used to record the amount of energy that is received at the top of the atmosphere. The conserved energy can then be calculated and be accounted. Different surface of the Earth will then have different rate of energy exchange. Therefore, the change in energy balance will affect the climate.

The loss of grassland have been measured by the proportion where it covers the globe. Grasslands that have been lost regionally will then be measured by various units such as kilometres square (km^2) and hectares (ha) on a larger scale. To understand further on how the grassland is affected by climate change and other factors, it can be measured with the annual changes in carbon stocks in grassland. Therefore,

$$\Delta C_{GG} = \Delta C_{GGLB} + \Delta C_{GGsoils} \qquad (3)$$

The annual change in carbon stocks is measures in tonnes of carbon per year and is derived from the sum of annual change in carbon stocks in living biomass (ΔC_{GGLB}) and annual change in carbon stocks in soils ($\Delta C_{GGsoils}$) in grassland. However, with this use of formula, to calculate the change of carbon stocks in different region will then take into account of the specific grassland type (i), the climatic zone (c) and the management regime (m). Since the ΔC_{GGLB} can be affected by different factors, regional grassland carbon stock can then be calculated with:

$$\Delta C_{GGLB(c, i, m)} = \left(\Delta B_{perennial} + \Delta B_{grasses}\right) \times CF \qquad (4)$$

CF is at the default of 0.5. where the change is the product of carbon fraction of dry matter (CF) to the sum of change in above- and belowground perennial woody biomass $\left(\Delta B_{perennial}\right)$ and below ground biomass of grasses ($\Delta B_{grasses}$).

Therefore, to accurately place the equation with the inclusion of the type of grassland, the climatic zone the grassland is in and the management regime that the grassland have been placed under:

$$C_{GG} = \left[\left(\Delta B_{perennial} + \Delta B_{grasses}\right) \times CF\right] + \Delta C_{GGsoils} \qquad (5)$$

Inventory system could also be set up to record clear data of the plants present, the climatic patterns and the management regime where animals that are grazing or being managed by humans efforts to conserve grassland.

4. Grasslands

Grassland is an area where various grasses dominate. Vegetation in grassland will grow no taller than the height of a shrub nor a tree. Little do people know that grasslands are one of the major ecosystems that covers close to one-third of Earth's terrestrial surface [6, 7]. In the last century, there is a decline in grasslands area worldwide to convert for arable land for production of animal feed crops and conversely, lack of management and abandonment [8]. Grasslands have been categorised as natural, semi-natural and improved grasslands [6, 9]. Natural grasslands are those that have been formed through processes that are related to the climate, fire and wildlife grazing. Semi-natural grasslands are those with human interventions. Scheduled grazing and hay-cutting are required for maintenance. Lastly, improved grasslands are pastures from ploughing and sowing agricultural varieties or non-native grasses with production value (**Figure 2**).

Grasslands across the globe are managed for a variety of purposes. They are valued for basic goods such as timber and water. They also provide forage, fishes and wildlife, and recreation resources. Grassland is a functional landscape that provides feed for grazing livestock. The landscape provided by the grassland has

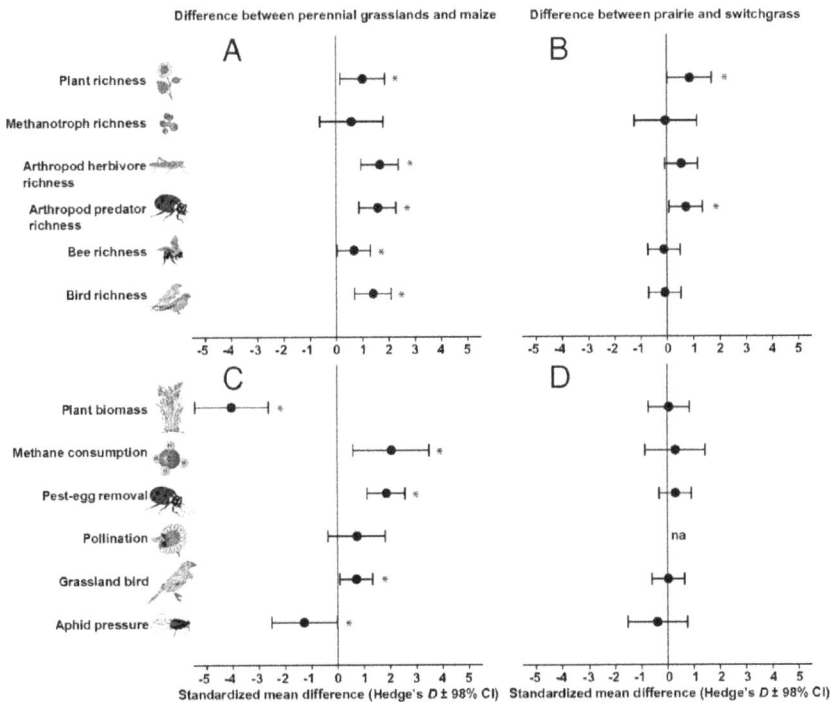

Figure 2.
Differences in richness and ecological processes were larger between the two perennial grasslands and maize than between prairie and switchgrass. Standardised effect sizes (Hedge's D) are shown for differences in richness and key ecological processes between grasslands and maize (A and C) (effect is difference between average of the two grasslands and maize) and prairie compared with switchgrass (B and D). Error bars show 98% confidence intervals. Asterisks indicate statistical significance at α = 0.02. Courtesy of Werling et al. [10].

Vector taxa	Vector group	Virus groups				Total	%
		Icosahedral particles RNA genome	Rod-shaped particles RNA genome	DNA genome	Enveloped particles RNA genome		
Hemiptera	Aphids	26	153[a]	13	5	197	28
	Whiteflies	—	13	115[b]	—	128	18
	Leafhoppers	8	—	15	3	26	4
	Planthoppers	10	4[c]	—	4	18	3
	Other hemiptera	—	8	5	—	13	2
Thysanoptera	Thrips	2	—	—	14	16	2
Coleoptera	Beetles	50	1	—	—	51	7
Acari	Mites	10	9	—	—	10	1
Nematoda	Nematodes	45	3	—	—	48	7
Mycota	Fungi	8	16	—	—	24	3
	No identified vectors	84	60	19	3[d]	166	24
	Total	253	268	167	30	697	
	%	33	39	24			

[a]Includes 110 virus species of the genus Potyvirus, family Potyviridae.
[b]Virus species of the genus Begomovirus, family Geminiviridae.
[c]These are all tenuiviruses that have multiple shapes.
[d]These viruses probably have insect vectors.
Courtesy of Hogenbout et al.

Table 1.
In Hogenbout et al. () study, they have discovered the types of genome that affects the diseases of grassland. The table then further explains that types of virus each insect group have been found as vector. However, this information may be limited to the area of experiment conducted and not in other continents [11].

often been perceived as free and limitless. The table below illustrated that the increase in plant richness will increased pollinators and diversity in the grasslands. With the increase in biodiversity, there was also an increase in pest-egg removal by the arthropod predator. Aphid pressure that is present in a grasslands have also decreased. All these meant that a healthy and well diversified grassland is able to strive and continue to prove various services indefinitely. This has highlighted the importance to conserve and improve grasslands health (**Table 1**).

The climate of grassland will be ideal for the growth of grasses only as low precipitation rate is not sufficient to sustain woody plants. Other maintaining factors of grasslands are fires and grazing animals. Grasses are well adapted to grow back after a fire as they have a complex root system and a resilient physiology. Grasses need not grow by seeds. Different part of the world will have different grassland climates. Therefore, they are differentiated by the Berkeley Biome Group.

With reference to the Berkeley Biome group, grasslands are categorised into two main types. These two types are differentiated according to their climate – tropical and temperate. Grasslands are sensitive to the change in climate as they have a strong seasonal climate. This suggests the possible changes that may occur to the characteristics of the grassland with long term exposure to climate changes. Other evidence also supports the hypothesis as there is a phenological and vegetation shifts[1] even before grasslands were impacted by climate change. Grasslands have provided a regulating services by providing climate regulation, carbon sequestration, erosion control, water regulation, air quality regulation, soil formation, pest control, waste treatment and pollination services. These natural environmental services are essential to keep the Earth's energy balance.

Climate change have since impact grasslands through increased seasonal, annual, minimum, and maximum temperature and the change in precipitation patterns [12]. Depending on the location of grassland, the climatic experiences can vary and theses variations include, increased temperatures, reduced rainfall and prolonged periods of drought. Grasslands are often bordered by forests, deserts seas and mountains. The change in earth's energy budget then have a vegetation shifts of either having rainforest encroaching into savanna and arid deserts being projected into arid grassland ecosystem. The slightest change in temperature, precipitation could alter the distribution, composition and the abundance of species in grassland. This would then result in the shift of products and services being provided. With the change in energy can also affect the geographical and elevational boundaries [13].

4.1 Adaptations of grasslands

Grassland then adapt to the change in climate by controlling on the opening of the stomata to optimise the balance between photosynthesis and transpiration. With the extended period of change in climatic condition, C3 plants[2] will no longer have the ability to flourish in such an environment and dormant C4 plants[3] seeds or any vegetative parts that is in the soil will flourish and take over the area. Hence, the vegetation shift. The shift in vegetation may not be just the end of the story. As the carbon dioxide levels in the atmosphere continue to rise, carbon, water and nitrogen cycles would also be affected in the grasslands. Gas exchange in the plants is a key player in these cycles. The reduced transpiration level will lead to a reduced mass flow from the soil to the roots and leaves, causing reduced nitrogen uptake and feedback to weaken photosynthetic capacity. The increase in carbon dioxide in the atmosphere has reacted to the change in climate by exhibiting their decreased nitrogen nutrition status [1]. Hence, the change in Earth's energy budget has impacted grassland by causing reduced stomatal conductance and significant reduction in yields.

For a single plant to strive in the environment, it requires certain criteria to grow. Without the criterion, the plant could have stunted growth, slow growth, discoloration and every other possibility. Apart from the observable morphology signs or poor health, they would also display poor yield. The application of chemical and/or organic fertilisers may not be effective as some plants just simply require concentrations that are readily available in the atmosphere. Thus, the instance stated above, where C4 plants are affected by the change have demonstrated that even if there is a shift in vegetation, if the area is deem inhabitable, the area will continue to remain dry and arid. Therefore, it is a relatively simple concept to comprehend – plants with poor health would then be highly susceptible to other diseases. The complexity of plant physiology is affected by the change due to the imbalance in earth's energy budget and to ensure the continuity of its own species.

---

[1] Vegetation shifts meant that plants that are C3, where carbon fixation takes place on a fix place and C4 plants where the carbon fixation takes place in both the mesophyll cells and in the bundle sheath cells.

[2] C3 plants utilise the Calvin cycle in the dark reaction of photosynthesis. Photosynthesis in these plants only take place when the stomata are open.

[3] C4 plants utilise the Hatch-slack pathway during dark reaction and have chloroplasts that are dimorphic. Photosynthesis in these plants will continue to occur even when the stomata are closed.

As mentioned earlier, climate change has affected grasslands with the change in temperature and precipitation patterns. In tropical grassland, the change of 1 Undercounter Temperature (uC) to 4 uC, will have grasslands experiencing increased aridity which reduces the productivity of soil organic carbon. It would also have plants experiencing increased water stress, therefore, altering the distribution pattern of grassland communities. There would also be decreased palatability of herbage and increased flammability. Drought tolerance species would then dominate and potentially lead to the extinction of other plant species. The carbon cycle in the tropical grassland would be affected as it will no longer be a carbon sink but it will be a source of emitting carbon dioxide. Grassland that is dominated by the C4 grasses will have enhanced biomass production because of the increase in soil moisture. When the grassland have low nitrogen availability, the response to elevated carbon dioxide will be suppressed. This would cause both long and short term effect. Impact of elevated carbon dioxide can be neglected during a short period of time with the increase in efficacy of nutrient-use and increased nutrient uptake due to higher root biomass at the elevated carbon dioxide.

Temperate grassland have similar changes experienced by the tropical grassland. Similar conditions such as increased drought due to the change in seasonal water regimes. The climate variability has then caused water stress. The water stress would cause reduced forage which then affects grazing livestock and also other animals that graze on the grassland. Grasslands have the capability to provide a buffer against climate variability [14]. The changes in temperate grassland will affect agriculture and grazing animals as compared to tropical grassland at the temperature variability is greater (**Figure 3**).

4.2 Impact of climate change

The model above was predicted on what will likely to happen in the year 2050. In this present day, the grassland ecosystem is already under serious threat. In the next 50 to 100 years, grasslands have been predicted to have lose between 8 to 10% of the vascular plant (the differences between these estimates are driven by different socio-economic scenario). The effect of the loss of biodiversity in grasslands and its effects on the carbon cycle would be an example of the synergism of global change drivers where biodiversity loss would constrain grassland ability to cope with the effect of other stressors such as climate change and ozone pollution [10]. The drastic change in

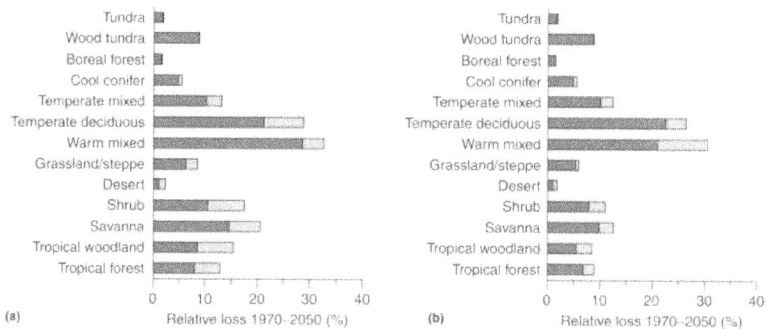

Figure 3.
Scenarios of biodiversity change for different biomes for the years 2020 and 2050. Bars represent relative losses of biodiversity of vascular plants through habitat loss for different biomes for two scenarios: (a) order from strength and (b) Adapting mosaic. Losses of biodiversity would occur when populations reach equilibrium with habitat available in 2050 and are relative to 1970 values. Darker bars represent scenarios for 2020 and lighter bars for 2050. Adapted from original Figure 10.6 in [15].

grassland would be impacted by human activities the greatest as the soil conditions are suitable for agriculture and the mild climate in the biome. However, this would then be driven by the increase in food security as the climate changes and the socio-economic status of natis widens. Hence, climate change do not only have direct threats on the grassland ecosystem, it also causes the change in mindset and management regimes.

5. Diseases mode of transmission

The ecosystem of grassland is diverse and complex. Apart from the climatic changes that the plants have to face, there are other threats that are threatening the peaceful existence of the C3 and C4 plants aside from being a food source to herbivores. Diseases in grassland can spread like wildfire depending on the mode of transmission. Several viruses or viroids could spread extensively in the field just by contact between healthy leaves. Viruses are spread systemically and can be transmitted through natural grafting. Root graft can also transmit viruses or even parasitic plants. Common transmission modes of plant diseases are often by vegetative propagation. There are also plant diseases that transmit through seed. For instance, sour-cherry yellows in the *Prunus spp*. The disease is caused by the Prune Dwarf Virus that causes young leave to have chlorotic yellow rings or mottle. The virus can be transmitted by infected pollen grains or infected seeds when pollinated by bees or during propagation process (**Table 2**).

5.1 Diseases transmission via vector

Plant diseases can be caused by various factors. Such factors could be abiotic such as nutrient deficiencies, soil compaction, salt injury or sun scorch. Biotic causes of plant disease transmission are caused by living organisms and they are collectively named as pathogens. Understanding pathogens life cycle, living requirements, movements and disease they carry can allow effective implementation of management regime to intervene the cyclative transmission.

Vector taxa	Vector species	Modes of transmission				Totals	%
		NPV[a]	SPV[b]	PCV[c]	PPV[d]		
Hemiptera	Aphids	161[e]	19	12	5	197	49.4
	Whiteflies	5	9	115[f]	—	129	32.3
	Leafhoppers	—	4	13	10	27	6.7
	Planthoppers	—	—	—	18	18	4.5
	Other hemiptera	2	9	1	—	12	3.0
Thysanoptera	Thrips	2	—	—	14	16	4.0
	Totals	170	41	141	47	399	
	%	42.6	10.3	35.3	11.8		

[a]NPV, nonpersistent stylet borne viruses.
[b]SPV, semipersistent foregut-borne viruses.
[c]PCV, persistent circulative (mostly nonpropagative) viruses.
[d]PPV, persistent propagative viruses.
[e]Includes 110 virus species of the genus Potyvirus, family Potyviridae.
[f]Virus species of the genus Begomovirus, family Geminiviridae.

Table 2.
The table illustrates the major group of insect that cause the virus spread through vector. The viruses have also been categorised to persistency in the environment [11].

5.1.1 Insects

Diseases in plants can also be transmitted via vectors. Aphids (28%) and whiteflies (18%) have been studied extensively over the years as they have been identified as common pathogenic vectors alongside with beetles (7%) and nematode (7%). However, the dense forests and every different area of land would always lead to a new discovery of a new species. In recent years, there is a new species of wasp, *Allorhogas gallifolia*. This wasp would make use of other wasp's gall as nests. Larvae that hatch would then feed on caterpillars that consume gall tissues. A caddisfly, *Potamophylax coronavirus*, has also been discovered. This moth has eggs and larvae that thrive in the environment near rivers and lakes. With new insects emerging, there are various study opportunities apart from just their life cycle. They can potentially be a reservoir host for all kinds of diseases to humans, animals and plants. The type of insects would often carry similar viroid.

Insects as vectors are relatively tricky to have a proper management to fully eradicate the population. This small, hardy population has found itself thriving through different ages of the earth by having the capability to populate through laying multiple eggs. These eggs laid by one are then sufficient to replace more than one adult in the population. With the rapid replication ability, insects are able to adapt to the changing environment through different mechanisms. Insects can transmit in a cyclative manner where the pathogen is ingested before passing on to the new plant host. Calculative plant pathogens will often induce physiological changes in their plant hosts. Vectors who then feed on the infected plant will then have behavioural changes to optimise the spread of pathogens to other plants. Their ability to invade a new area is dependent on the insects' ability to adapt to the environment besides the food availability.

5.1.2 Herbivore as vector

Herbivores are essential to the plant communities as grazing removes substantial quantities of biomass and promote plant species diversity [16]. In addition to herbivore vertebrate, insects have been shown to promote species richness by feeding competitive dominants. An instance to display such phenomena would be molluscs. They are a major group of invertebrate in temperate grasslands. These principal forb feeders would contribute negative impacts on species richness. These invertebrates are more well studied compared to soil-borne fungus. The complexity of the pathogen physiology is hard to comprehend as the environment will cause compensatory and additive interactions. Apart from infecting plants. Pathogens have displayed that they have the ability to increase plant susceptibility to herbivores to feed on the plant. Pathogens could also decrease it's susceptibility as it makes the plant less palatable to grazing animals.

5.2 Soil-borne pathogens

Besides insects as vectors for pathogens, they could be soil-borne as well. They are capable of spreading via swimming spores of primitive and soil-inhibiting pathogenic fungi. Fungal pathogens have been categorised into three functional groups: biological controllers, ecosystem regulators and species participating in organic matter decomposition and compound transformations. Fungal pathogens are dispersed by spores. Their successful inhabitants of soil is accredited to its high plasticity and their ability to adopt various forms in response to adverse or unfavourable conditions [5]. Biological controllers can improve soil health by regulating diseases, pests, and the growth of other organisms. Fungi as ecosystem regulators are responsible for soil structure formation and medication. This will enhance the habitat of other organisms through the regulation of the dynamic aspect of the physiological processes in the soil environment.

Infected grasslands were observed to have increased species richness. However, This would have a negative impact on the dominant species in the grassland. In addition, the affected grassland will have decreased biomass. In a study by Allan et al. (2010), the biomass of the grassland will increase in an increasing exponential manner over the years. Fungal pathogens are not harmful to all. Despite the harm the bring to some of the species, fungi actively participate in nitrogen fixation, biological control of root pathogens, production of hormone and protection against drought. For instance, fungi can be beneficial to some leguminous crop via the improvement of plant uptake of nutrients and provide some form of protection to pathogens.

Nevertheless, bacteria that cause diseases in plants. Bacteria could be transmitted via the similar route as viruses by having physical contact on the health leaves or introducing plant materials that have bacteria. The simple transmission mode carries a huge load of impact on the environment as it alters the plants' physiology.

6. Grassland diseases

Plant diseases are generally cause by three categories – (i) microscopic organisms like the fungi, bacteria and nematodes, where they have the ability to penetrate and infect more than one type of host, (ii) the sub-microscopic organisms such as viruses as they enter and infect the plant host systematically, (iii) parasitic higher plant that feed off their host.

Grassland diseases have been commonly affected by rust or having a weak rooting system. Disease have also been mentioned previously that transmission could be affected by vectors. Vector survival in the environment will then be crucial to understand the potential and possibility of having new diseases emerging. Studies have managed to display and explain different sources of colonisation for aboveground and belowground microbial communities and different drivers of community assembly. Grassland diseases are also dependent on the type of grasses. Diseases have affected the three main categories – grasses, legumes and cereal crops. Generally, they are all in a similar situation where they have insects as vectors (mainly aphids, beetles and soil-borne larvae), having mycosis and bacteriosis. Mycosis are caused by fungal pathogens where they destroy the plant tissue directly or through the potent toxins, this could be fatal to the host plant and can lead to ergotism in animals when they consume. Bacteria in plants can cause different kind of symptoms such as galls, overgrowth, wilts, leaf spots, leaf specks and blights. Occasionally, soft rots, scabs and cankers can be observed. In comparison to plant viruses, plant bacteria are not invasive and plant often occur as secondary infection through a vector.

6.1 Microscopic organisms in grassland

The soil biota often have bacteria, algae, fungi and soil invertebrates. The diversity of these microorganisms have been underestimated and under-researched. Microscopic organisms can have a symbiotic or a mutualistic relationship with the plant depending on the nature of the micro-organisms. Biological soil crusts have a biological community that is living on the soil surface. They perform several vital functions in grassland such as retention of moisture, stabilises surface soils, enriching soils with nitrogen and carbon, and even providing a favourable microclimate for seed germination The mosses and lichens will have rhizines, the gelatinous sheathe of mobile cyanobacteria, and fungal hyphae can bind surface soil particles to reduce soil and wind erosion. A well-developed biological soil crusts is an important factor in successful post-fire revegetation as they retain the integrity of the soil surface to provide a sanctuary for seeds propagules.

6.1.1 Bacteria

In the soil biota, the grassland have a diversity of bacteria. Bacterial communities were more spatially structured than fungal communities. Those bacteria can be an advantage to the grassland. Similarly, they can also be harmful to the host. *Rhyzobium spp.* are bacteria that live in the nodules of the roots of the leguminous plants.

In a study in Eastern of Czech Republic, bacterial sequences belonged mainly to Proteobacteria (50%) and Actinobacteria (20%). In shoots, the most abundant bacterial genera were *Vibrio*, *Pantoea* and *Pseudomonas*, all of which belong to Gamma-proteobacteria. In roots, the most abundant bacterial genera were *Vibrio*, *Chthoniobacter* (phylum Verrucomicrobia) and *Paeniglutamicibacter* (phylum Actinobacteria). In soil, the most abundant bacterial genera were *Chthoniobacter*, *Gaiella* (phylum Actinobacteria) and *Paenibacillus* (phylum Firmicutes). With such different bacteria that are found above and below ground, it supports the statement where different environment drivers can drive the formation of different colonies to be form or even the presence and/or the absence of some bacteria above and underground.

Every grassland may have different colonies of bacteria present. Those that are harmful to the host are mainly *Xanthomonas spp.*, *Pseudomonas spp.* and *Aphrodes spp.*

Xanthomonas translucens pv *graminis* is infamous for bacterial wilt in forage grasses that have reportedly caused an outbreak in Europe, Australasia and United States of America (USA). The infection starts from a wound site and will eventually lead to necrosis starting from the infected site. The progression would be made towards the leaf base or the host plant. When the bacteria reached the vascular tissue of the host, the bacteria will colonise rapidly throughout the plant causing wilting of leaves. The plant will eventually be killed within a number of days. Severe yield loss have been experienced in the temperate region. However, *translucens* pv *graminis* is not the only specie in *Xanthomonas* to have caused such massive disruption.

The other three species of *Xanthomonas* are *translucens* pv *arrhenatheri*, *translucens* pv. *poae* and *translucens* pv. *phlei*. These species have presented distinct host adaptations to the plant species and have been successfully isolated. The strains display low genetic diversity. These host specialised parthovar strains will allow insight into distinct virulence factors where host-specific adaptation at molecular level with reference to *Xanthomonas translucens* pv *graminis* in future studies.

6.1.2 Fungus in grassland

Fungal pathogens are strongly influenced by the diversity and composition of the plant community. As such, they have a return effect on plant growth through mutualism, pathogenicity and their effect on nutrient cycling and availability.

Grassland grasses have been observed to be commonly affected by crown rust. Crown rust is common on swards often when bulk of material has been built up for autumn grazing or a late silage cut. The disease is more prevalent when grass depletes its nutrients. The increased temperature have then encouraged crown rust to have increased occurrence. The severity of the infection can result in reduced yield and the palatability is adversely affected. The disease is favoured by warm, moist weather with tropical temperatures. The extreme differences in daily temperature are even more highly favoured. The transmission of disease is through wind and precipitation.

Crown rust were affected by a plant pathogen *Puccinia coronata*. This fungus have affected plants like oats, barley and most specifically ryegrasses. The orange pustules on the leaf blade produces uredospores that could spread long distances to other plants in the grassland. Black pustules will then produce teliospores and will remain on plant debris over winter. The spores will then stay dormant till spring. Teliospore will then later produce basidiospores that infect secondary hosts. Basidiospores will then further produce aeciospore that will repeat the infection process.

Plant disease mechanisms in specialist and generalist pathogens can promote unwanted diversity of diseases if the dominant species that is susceptible is present in the community. This has a similar concept as the maintenance of the coexistence between herbivores and the plant communities. In a community that have pathogens that attack on less competitive species will cause and adverse effect.

6.2 Sub-microscopic organisms in grassland

Viruses have always exits and have remain its unpredictable virulence, creating havoc in grassland. The most Barley and Cereal Yellow dwarf virus (BYDV) have been one of the most complex and threatening to both food security and the ability for the plant of this species to continue to survive. This virus have then play an important role in the competitive dynamics of native and invasive grasses in non-managed system [17]. As the name of the virus suggests, it is highly infectious among barley and cereal. However, in recent studies, the virus have been discovered to have infect invasive species, *Venenata dubia*, in grassland habitats. Aphids have been positively identified to have been the vector of this virus. However, no two aphids are the same. Non-colonising aphids have been suggested to be responsible for the expensive spread of the virus [17].

Apart from BYDV, the cocksfoot streak virus (CSV) has been at a rise as the virus turn pastures into hay-like texture. CSV is aggressive and have reduce the quality of hay. Plants that were infected have also reduced ability to withstand frequent defoliation [18]. The virulence of CSV is not as aggressive as BYDV as progenies of infected plant do not have strains of CSV unlike BYDV.

7. Climate impact on vectors

Grassland diseases are highly affected by the availability of the mode of transmission in the environment. The knowledge on entomology is required to understand the point of interference in the life cycle in order to have successful management strategies.

7.1 Aphids

Aphids belong to a superfamily of Aphidoidea, which belongs to the Hemipteran sternorrchyna with whiteflies, jumping plant lice, scale insects and mealybugs. This superfamily is then separated into two sub groups of primitive "aphids" and a group of new world aphids [19].

The studies on aphids' ability to transmit viruses have become more complex as they are capable of switching between sexual and parthenogenetic reproduction. Aphids are a vector of Ribonucleic Acid (RNA) grass pathogens and such viruses include barley and cereal yellow dwarf viruses. The severity of grassland diseases are dependent on the areas located. For instance, in California, United States of America (USA), the pathogen-mediated invasion in grasslands is the result of competition between native and exotic plants where aphids have higher fecundity on exotic plants compared to that of the natives [20]. This factor has potentially led to the increase in pathogens transmission rates throughout the community. The life cycle has then been studied closely to understand and in hopes to discover a point of interference which would break the reproduction rate.

A single host plant species is often observed to be utilised throughout the year. In response to the decreasing daylight, sexual morphs are produced in the fall. Genetically recombinant eggs that are reproduced by the male and his oviparae[4] mate would overwinter on the host plant and often experience a high mortality rate. Fundatrix[5] that emerges from the eggs in spring will proceed to reproduce to live births parthenogenetically. These nymphs would be viviparae[6] and will continue the lifecycle in summer. The parthenogenetic mode of reproduction has ensured a rapid population build-up by ensuring that there are eggs available on the plant year round. The rapid increase in aphids population in a single host plant could quickly lead up to overcrowding. This would allow the future offspring of the aphids would be switched to those with wings to have efficient dispersal of feeding opportunity and ensures the genetic survival.

Aphids (49.4%) are transmitters for the majority of the mosaic virus and leafhoppers (6.7%) are transmitters for yellow-type viruses. There are many other insects of interest such as whiteflies (32.3%), thrips (4.0%), mealybugs, plant hoppers (4.5%), grasshoppers, scales and beetles. Aphids are sap-sucking insects and have piercing, sucking mouthparts. They strive generally well in regions with cold winters. The use of their mouthparts include a needle-like stylet that assist aphids to have access and feed on the contents of plant cells. The insect's feeding habits will weaken the plant and cause metabolic imbalance. In addition, aphids secrete honeydew. This is an ideal medium for a variety of fungi to populate. As such, sunlight would be blocked out as the fungi populate, building a barrier for the plant to photosynthesize. During the process of feeding, their stylet has created a point of entry for the pathogens to enter the system of the plant host. The plant, if infected with secondary infection, would then be infected and display disease symptoms.

Aphids have been covered in a relatively large proportion in this chapter. This insect has eggs that are cold-hardy to survive winter. The efficacy to have population build-up is only possible when the temperature is optimum. Every species of aphids have different optimal temperatures. However, the minimum range is said to be at 4 degree Celsius. *Acrythosiphon pisum* is an aphid that reproduction is dependent on the temperature. The overall increase in global temperature by 2 degree Celsius would have an approximation of generations increased from 18 to 23 generations per year (based on a study in the United Kingdom). However, the generation time of a female differs between species and this could potentially be shorten by the decrease in temperature due to global warming. France has a mean temperature of 10 degree Celsius (in the north) and 15 degree Celsius (in the south)

---

[4] Oviparity refers to female that produce eggs, not live young.

[5] Fundatrix is a viviparous parthenogenetic winged or wingless female aphid produced on the primary host plant from an overwintering fertilised egg.

[6] Viviparity meant that the female bringing forth live young which have developed inside the body of the parent.

which place the aphids in suboptimal temperature conditions. However, this is an alarming increase for entomologists to study aphids further. Apart from the temperature being favourable to the rate of reproduction. The temperature increase is also favouring the mobility of the aphids. The winged aphids have a threshold of 13 to 16 degree Celsius and an upper threshold of approximately 31 degree Celsius [18].

7.2 Soil dwelling organisms

Apart from aphids that are attacking above ground, there are also vectors attacking below ground. Larvae of *Cerapteryx graminis*, a moth from the *Lepidoptera*, are soil-dwelling and the larvae can cause *Charaeas graminis* on grasses. Larvae of *Tholera decimalis* Poda (from *Lepidoptera*) are soil-dwelling and can cause disease to a plant by feeding on its roots. Other larvae that are soil-dwelling can cause large amounts of damage to the plant as the larvae mainly feed on the roots of the host plant and some adults may continue to dwell in the same plant causing more harm. Larvae feeding at the root system may invite secondary infection causing more complications. Nematodes and soil-dwelling borers are also a vector for infection in plants as they could create entry for bacteria to cause further complications to plant health.

These soil-dwelling organisms will be impacted by the decrease in moisture in soil (regions where desertification occurs). Increase in flooding will also be a concern to these organisms as they may not survive if the soil moisture increases too drastically. Their living conditions are also affected by the temperature. The adaptation to the changing climate is similar to that of other insects living aboveground.

7.3 Soil bacteria

Bacteria can be transmitted naturally through exudation out of the host plant and when contact is made between injured plants, they can infect the plant through the wound site. Insects that come in contact with the exudates that infected host plants produce, they can also transmit to other plants as secondary infection. Insects are often attracted to the sugars in the bacterial exudates. During the process of consumption, the mouthparts of the insects will then carry the strains of bacteria. Upon travelling and feeding on other plant, they will create an entry for these bacteria they carry and the plant will now be infected.

Bacteria being microscopic organisms will be sensitive to the change in environment conditions. Depending on the types of bacteria, some may strive in the areas of higher temperature like the *Xanthomonas spp.* are at advantage but not for *Puccinia spp.* The virulence of the bacteria have been studied to have been affected by the change in temperature. *Agrobacterium* strains have their virulence gene amplified as temperature increases but they will have a loss of phosphorylation activity [21]. *Pseudomonas* have increased production of phytotoxins as there is an increase in temperature to maintain its virulence. Therefore, the change in climate will affect the physiological functions of the bacteria differently and ensure the continuum of bacteria in the environment. The effect of bacterial virulence of some effectors may become apparent under specific environment conditions such as humidity.

7.4 Soil fungus

Soil fungus has different roles in the soil which then serves different ecosystem services. They are also bioindicators of soil health. However, as mentioned earlier, crown rust is a genus of fungus (*Puccinia spp.*). Some of the *Puccinia spp.* are considered as parasites of plants. The presence of harmful plant pathogens indicates poor soil quality. The factors that cause changing soil fungal biodiversity are mainly due to the management practices, chemical fertilisation, application of herbicides and fungicides, biochemical amendments of the soil, soil degradation, soil contaminants and soil properties such as salinity and drought conditions.

Global warming can influence the host plant associations through alteration of interactions between plant and mycorrhizal fungi. This group of fungi have the role of having direct influence on individual plant function and the indirect impact processes such as plant dispersal and community interactions. However, to mediate and to survive, they have ways to mediate the current changes of climate. Such methods involve varying in hyphal exploration type liked to root density [5, 22]. Climate change does not seem all that bad when the essential fungus in the soil required are still able to survive.

The change in climate has created new environmental pressures that results in novel fungus diseases. The effect of climate change on the emergence and re-emergence of fungal pathogens have raised concerns on food security, human and animal health, and wildlife extinction due to the report worldwide. There are new virulent fungal lineages with adaptations emerging and they have been suggested to have evolved alongside with the increased pressure of climate change. One such fungus is the *Puccinia striiformis*, commonly known as rust fungus (same genus to the current crown rust pathogen). Stripe rust has affected wheat crops worldwide. There were records that indicate the preference for cooler regions but has recently invaded to warmer regions. The ability to disperse to warmer regions has allowed the emergence of three novel strains. These strains have been described as being more aggressive with increased thermotolerant [23]. The spread of novel strains have been hypothesised for having the ability to replace older strains and expanding the spread of disease. Through microsatellite genotyping and virulence phenotyping on the novel strains, it has been demonstrated that the evolution can potentially be ongoing alongside with the change in climate.

Another fungal concern would be the emergence of *Fusarium* head blight in wheat and other cereal crops [14]. The infection can reduce crop yield and quality. Thus, threatening food security. Outbreaks have been reported specifically with years that experience warmer and humid weather. The economic loss during outbreaks could be up to 75 percent [4]. The shift in temperature due to climate change have allowed the fungi to be more aggressive and able to expand the spread of territory. The change of favourable weather has been observed in two species of the *Fusarium* genus. *Fusarium graminearum* and *Fusarium culmorum* are two of the species that display prominent change towards the shift in temperature over the regions. They have very contrasting weather preferences. Thus, these fungus can expand through larger areas that do not adapt. The increase in environmental stress due to the change in climate have also evidently shown some of the species to react by producing more mycotoxins. As such this has a rising concern not only to food security but also human and animal health.

Apart from the changes that the fungi have evolved to ensure the survival of its kind, the spread of spores have then been extensive through the rising disastrous events. Frequent flooding and strong winds causing dust storms are two such extensive transmission methods. Soil-borne fungal pathogens have been speculated to have increased frequency or range due to climate change [6]. They are found out of their normal range and at times can be challenging to have a first diagnosis [24].

8. Future perspective

There might be results that the resistant plant type is achieving ideal suppression of damage done. However, all living things have the ability to change and adapt to the environment they live in. Plants that as antibiosis may achieve ideal results when planted in Region A. However, when planted in Region B and C, the result may vary. Assuming that the soil conditions in all three regions are the same. However, abiotic factors cannot be controlled and that may be the factor that causes the difference in results. Therefore, grassland management has to be very specific to a particular location and changes that occur through the years can be used to study closely to have a more effective management plan.

Disease in grassland have been affected by the Earth's energy imbalance due to the change in living environment and transmission mode. The change in climate have affected the population of the vectors. Having vectors in the environment is essential for transmission mode in the ecosystem. The reduced availability vectors in the population will ideally have a decreased in the extensiveness of the spread of disease. However, if the disease, in particular consideration to viroids, where to mutate, the mode of transmission could change to either, air-borne or even have a longer dormancy capability to ensure sustainability of its existence.

8.1 Innovative management

The different kinds of vectors will require different methods of surviving as the climate changes. The increase in certain greenhouse gases in the atmosphere makes it complex to understand the change that the vectors are going through. However, there are a few significant points that could be brought across in this chapter. Vectors such as insects have expanded their distribution to regions where it will be more habitable to them. This can be supported by the pink bollworm (*Pectinophora gossypiella*), an infamous cotton pest that has expanded towards the central of California and away from the South. Other vectors such as the Olive fly (*Bactrocera oleae*) have demonstrated migration behaviour. They would travel southwards during winter to experience summer in other areas. However, this would increase competition of insects for the availability of food for the population. This will potentially lead the insects to have a change in diet if available and adaptable.

Apart from migration behaviour, vectors that have remained have increased in overwintering survival. They will produce eggs that are more hardy to withstand the change in temperature and environmental damages. Insects have also adapted to migrate as mentioned earlier. Vectors that are freeze-tolerant have physiological adaptation to be diapause[7]. They could be obligate or facultative. Regardless of which they are, the insects will be hormonally mediated to a state of having low metabolic activity. This will suppress development, suspend activities and increase resistance to adverse environmental factors change. Insects will also display aestivation or hibernation. The ability for the insect to synchronise with the changing environment will be the most ideal situation where the expansion and the spread of diseases are still highly plausible.

The change in ambient temperature have accelerated reproduction rates. This has caused an increase in population size. As such, it can lead to the number of species having dynamic equilibrium. To understand the phenological shifts caused by climate variability, it has been measured with growing degree days (GDD). The GDD will then aid in determining the minimum and maximum temperature threshold. Insects of multivoltine, such as aphids, are at the advantage of the rising temperature. Increase in 2 degree Celsius in temperature could have an estimate of additional five generations. Other insects demonstrated having earlier flight as the ambient temperature increases.

---

[7] Diapause is an adaptive trait that plays an important function in the seasonal regulation of insect life cycles and is influenced by environmental factors.

The change in climate has also brought about the change in precipitation patterns. High rainfall will have insects such as aphids being washed off and will decrease the opportunity of the insect or pathogens overwintering. However, the insects can migrate further up the soil horizons or deep down. Soil-dwelling wireworms have adapted to the change in precipitation pattern by populating on the upper soil horizon and migrate as they grow as an adult.

8.2 Potential limitations

The cruciality of understanding vectors in plant diseases in relation to that of climate change is complex and underestimated. The importance of vectors' movement and traceability have yet to be identified clearly as they are showing signs of evolution with rapid reproduction rate. As such, plant diseases can be said to have spread as rapidly as the vectors expand their area of infection. Plant diseases cause secondary infections which are mainly facilitated by insects to allow entry to a pathogen by creating a wound site on the plant organ regardless of it being above or underground. Pathogens that are vectored by insects can also overcome survival to adverse environment factors through the maintenance of over-seasoning in the body of the insect.

There are different management methods innovated in order to suppress the damage incurred. Having a plant that is resistant to specific insects or pathogens is an innovative way or management. For example, antibiosis in host plant resistance is a primary mechanism that works against aphids [17]. The process occurs at the utilisation phase of the interaction between the plant and the insect. It is the result of action of plant-biochemicals in the biological processes of herbivorous insects. Antibiosis would then be expressed in terms of larval mortality, decreased larval and pupal weights, prolonged larval and pupal development, reduced fecundity, prolonged generation time and overall effect on insect survival and development.

9. Conclusion

Climate change has created a new ecological niche and opportunities are provided for vectors to continuously expand their geographic region. Hence, the migrating behaviour. Microscopic plant pathogens and vectors who spend most of their life underground have a comparatively greater advantage of surviving climate change as soil is a thermal insulating medium, buffering temperature change and reducing impact. Apart from the focus of climate change being the increase in global temperature, the change in climate is contributed by human activities where the atmospheric composition changes. The significant gas that all scientists are studying is the carbon dioxide concentration. The increase in carbon dioxide concentration has driven vegetation shift. However, it has also driven the susceptibility of pathogens of these vegetation. High carbon dioxide levels can encourage plant growth. However, it will encourage the feeding for insects as vegetation increases in palatability.

The change in the ecosystem is tied in closely to that of insects and pathogens. Therefore, the change in climate will strongly affect the survival of the vectors of diseases rather than the diseases itself. This can be supported by the expansion of diseases as the vectors expand their movement through migration. The behavioural changes in vectors are significant and as they strive and adapt to the change in climate, so will the plant diseases in grassland continue to cause more damage.

Author details

Ang Jia Wei Germaine
The University of Queensland, Queensland, Australia

*Address all correspondence to: gxrmainx@gmail.com

IntechOpen

References

[1] Emadodin I, Corral DEF, Reinsch T, Klub C, Taube F. Climate Change Effects on Temperate Grassland and Its Implication for Forage Production: A Case Study from Northern Germany. Agriculture. 2021;**11**(3):1-17. DOI: 10.3390/agriculture11030232

[2] Clark MF, Christensen MJ. Some Observations on an Aphid-borne Virus Disease of Ryegrass in New Zealand. New Zealand Journal of Agricultural Research. 2012;**15**(1):179-183. DOI: 10.1080/00288233.1972.10421292

[3] Craine JM, Ocheltree TW, Nippert JB, Towne EG, Skibbe AM, Kembel SW, et al. Global diversity of drought tolerance and grassland climate-change resilience. Nature Climate Change. 2013;**3**:63-67. DOI: 10.1038/NCLIMATE1634

[4] Ingwell LL, Lacroix C, Rhoades PR, Bosque-Pérez NA. Virus infection in an endangered grassland habitat. International plant virus epidemiology symposium. 13; 165p. DOI: hal-02740486

[5] Leplat J, Friberg H, Abid M, Steinberg C. Survival of Fusarium graminearum, the causal agent of Fusarium head blight. A review Agron Sustain Develop. 2013;**33**(1):97-111. DOI: 10.1007/s13593-012-0098-5. hal01201382f

[6] Kölliker R, Krähenbühl R, Schubiger FX and Widmer F. Genetic Diversity and Pathogenicity of the Grass pathogen *Xanthomonas translucens* pv. *graminis*. Molecular Breeding of Forage and Turf. 2004; p. 53-59. DOI: 10.1007/1-4020-2591-2_5

[7] Suttie JM, Reynolds SG, Batello C. Grasslands of the World. Rome, Italy: FAO; 2005

[8] Queiroz C, Beilin R, Folke C, Lindborg R. Farmland abandonment: Threat or opportunity for biodiversity conservation? Frontiers in Ecology and the Environment. 2014;**12**:288-296

[9] Borer TE, Adams VT, Engler GA, Al A, Schumann CB, Seabloom EW. Aphid Fecundity and grassland invasion: Invader life history is the key. Ecological Application. 2009;**19**(5): 1187-1196. DOI: 10.1890/08-1205.1

[10] Werling BP, Dickson TL, Isaacs R, Gaines H, Gratton C, Gross KL, et al. Perennial grasslands enhance biodiversity and multiple ecosystem services in bioenergy landscapes. PNAS. 2014;**111**(14):1652-1657. DOI: 10.1073/pnas.1309492111

[11] Hogenbout SA, Ammar ED, Whitfield AE, Redinbaugh MG. Insect vector interactions with persistenetly transmitted viruses. Annual Review of Phytopathology. 2008;**46**:327-359. DOI: 10.1146/annurev.phyto.022508.092135

[12] Sala OE, Vivanco L, Flombaum P. Grassland Ecosystem. Encyclopaedia of Biodiversity. 2013;**4**:1-7. DOI: 10.1016/B978-0-12-384719-5.00259-81

[13] Lemaire G, Hodgson J, Chabbi A, editors. Grassland productivity and ecosystem services. Wallingford, UK: CABI; 2011

[14] Catherall PL. The Significance of Virus Diseases for the Productivity of Grassland. Journal of the British Grassland Society. 1966;**21**(2):116-122. DOI: 10.1111/j.1365-2494.1966.tb00458.x

[15] Sala OE, van Vuuren D, Pereira H, et al. Biodiversityacross scenarios. In: Carpenter SR, Pingali PL, Bennett EM, and Zure kM (eds.). Ecosystems and Human Well-Being: Scenarios. Washington, DC: Island Press; 2005. pp. 375-408

[16] Augustine DJ, McNaughton SJ. Ungulate effects on the functional

species composition of plant communities: herbivore selectivity and plant tolerance. Journal of Wildlife Management. 1998;**62**:1165-1183

[17] Frac M, Hannula SE, Bełka M, Jedryczka M. Fungal Biodiversity and Their Role in Soil Health. Front. Microbiol. 2018;**9**:707. DOI: 10.3389/fmicb.2018.00707

[18] Dweba C, Figlan S, Shimelis H, Motaung T, Sydenham S, Mwadzingeni L, et al. Fusarium head blight of wheat: Pathogenesis and control strategies. Crop Prot. 2017;**91**:114-122. DOI: 10.1016/j.cropro.2016.10.002

[19] Irwin ME, Kampmeier GE, Weisser WW. Aphid movement: process and consequences. In: van Emden HF, Harrington R, editors. Aphids as Crop Pests, CABI, UK; 2007. pp. 153-186

[20] Bingham MA, Biondini M. Mycorrhizal hyphal length as a function of plant community richness and composition in restored northern tallgrass prairies (USA). Rangeland Ecology & Management. 2009;**62**:60-67. DOI: 10.2111/08-088

[21] Mahlman JD. Uncertainties in projections of human-caused climate warming. Science. 1997;**278**(5342): 1416-1417. DOI: 10.1126/science.278.5342.1416

[22] Bengtsson J, Bullock JM, Egoh B, Everson C, Everson T, O'Connor T, O'Farrell PJ, Smith HG, LindBorg R. Grasslands – more important for ecosystem services than you might think. Ecosphere. 2019;**10**(2):1-20. DOI: e02582. 10.1002/ecs2.2582

[23] Bullock JM et al. Chapter 6: Semi-natural grasslands. Pages 161-196 in UK NEA. The UK National Ecosystem Assessment. Cambridge, UK: UNEP-WCMC; 2011

[24] Liang S, Wang D, He T, Yu Y. Remote Sensing of Earth's Energy Budget: Synthesis and Review. International Journal of Digital Earth. 2019;**12**(7):737-780. DOI: 10.1080/17538947.2019.1597189

Section 2

Underutilized Grasses Production Potential

Chapter 4

Underutilized Grasses Production: New Evolving Perspectives

Muhammad Aamir Iqbal, Sadaf Khalid, Raees Ahmed,
Muhammad Zubair Khan, Nagina Rafique, Raina Ijaz,
Saira Ishaq, Muhammad Jamil, Aqeel Ahmad,
Amjad Shahzad Gondal, Muhammad Imran,
Junaid Rahim and Umar Ayaz Aslam Sheikh

Abstract

Globally, over-reliance on major food crops (wheat, rice and maize) has led to food basket's shrinking, while climate change, environmental pollution and deteriorating soil fertility demand the cultivation of less exhaustive but nutritious grasses. Unlike neglected grasses (grass species restricted to their centres of origin and only grown at the subsistence level), many underutilized grasses (grass species whose yield or usability potential remains unrealized) are resistant and resilient to abiotic stresses and have multiple uses including food (*Coix lacryma-jobi*), feed (*Eragrostis amabilis* and *Cynodon dactylon*), esthetic value (*Miscanthus sinensis* and *Imperata cylindrica*), renewable energy production (*Spartina pectinata* and *Andropogon gerardii Vitman*) and contribution to ecosystem services (*Saccharum spontaneum*). Lack of agricultural market globalization, urbanization and prevalence of large commercial enterprises that favor major grasses trade, improved communication means that promoted specialization in favor of established crops, scant planting material of underutilized grasses and fewer research on their production technology and products development are the prime challenges posed to underutilized grasses promotion. Integration of agronomic research with novel plant protection measures and plant breeding and molecular genetics approaches for developing biotic and abiotic stresses tolerant cultivars along with the development of commercially attractive food products hold the future key for promoting underutilized grasses for supplanting food security and sustainably multiplying economic outcomes.

Keywords: agronomy and food sciences, plant protection and breeding, new crops, entomology, plant pathology

1. Introduction

Grasses biodiversity constitutes one of the critical primary sources for securing sustainable supplies of food, feed, fiber, medicines, aromatic stuffs and shelter [1–3]. Globally, humans have put to use a very limited number (less than one-third) of plant species from the recognized pool of species which diversified generations of varying cultures have been aware of for multiple uses. The origins and regions of diversification for numerous underutilized grasses have been investigated recently, but information pertaining to genetic diversity and agro-botanical traits of many species having local pertinence has remained scant. One of the underlying reasons for over-focussed staple crops might be attributed to overwhelming reliance on prime food crops which has led to food basket's shrinking across the globe [4, 5]. This phenomenon finds its roots in the simplification and intensification of agricultural production systems. These have historically favored major grasses over others owing to their comparative and competitive advantages regarding successful production in a wider range of pedo-climatic conditions, feasible cultivation requirements, grower-friendly processing, economical storability, unmatched nutritional properties, high market demand, superior revenue generation and preferable taste [6–8].

It is also pertinent to mention that the simplification process of agricultural production systems has abruptly lowered the quality of agricultural produces over time. However, this approach has reduced the risks of complete crop failure and multiplied successful harvests opportunities. This in turn has boosted survival human's survival through limited but sufficiently produced yields by major grasses. The commercial-oriented farming systems focussing major grasses have caused serious decline in intra and interspecific diversity of crops. In addition, other dis-advantages of over-emphasis on major grasses include higher vulnerability among growers and end-users, for whom grasses diversity have become survival necessity rather than a matter of choice due to changing climate and crop failures owing to a bunch of biotic and abiotic stresses [9, 10].

2. What are underutilized grasses?

Among agriculture terminologies, perhaps underutilized term has given rise to wider discussions and debates since long. It is normally applied to grass species whose yield or usability potential remains unrealized. However, this definition might be declared inconclusive owing to missing information regarding underuti-lization in what sort of geographical regions, cultures and economic feasibilities. Thus, using this term inevitably needs a clarification to explicitly describe the exact meanings and applicability of the term. For instance, with respect to geographical implications of the term, a grass species might be underutilized in some regions compared to others. Regarding economic applicability, some grass species might constitute as vital component of masses daily diet, but these may largely remain underutilized in other regions owing to poor marketing conditions and lower economic turnout for growers. As far as time factor is concerned, dynamic market-ing systems might improve the degree of underuse due to vigorous attention in few regions while the same species could continue to witness poor marketing owing to lesser attention of growers and researchers in some regions. For example, hulled wheat which represents the collective name of *Triticum monococcum*, *Triticum dicoccum* and *Triticum spelta* has attained the status of a speciality crop in Italy along with many other European countries, whereby numerous ex situ and in situ techniques for its conservation are being developed through integrated research efforts. However, the same grass constitutes the status of a life support crop in Turkey's remote areas. Few underutilized grasses are being marketed as new crops. However, the fact is that different commercial companies and researchers have recently started working on them for boosting their productivity and nutritional value. The reality is that local populations have used underutilized grass species for generations while these remained unattended historically. It is worth mentioning that locally based knowledge and traditional uses of underutilized grasses at limited scale have contributed to portray such underutilized grasses as new crops [1, 4, 11].

3. Differentiation of underutilized and neglected grasses

The underutilized grasses might be defined as grasses that were traditionally grown widely in localized production systems but now their cultivation and use have seriously decline owing to cultural, genetic, agronomic, climatic, economics, globalization and market related factors. Their cultivation and consumption have reduced significantly for their being non-competitive with staple grasses in the same agro-environmental conditions. The net result is the eroding of the grass genetic pool which has narrowed down the choice of crops for improvement as well as adaptation under changing climatic scenarios.

Contrastingly, neglected grasses tend to remain restricted in their centres of origin and are primarily grown by local farmers at the subsistence level. It may be noted that few grass species could be globally; however, tend to prevail in few special niches within local ecology and traditional agricultural systems. Thus, these types of grasses have continued to be grown on limited scale under socio-cultural choices, however these have remained inadequately characterized and historically neglected by researchers, agronomists and conservationists [12].

4. Agro-botanical superiority and multipurpose utilization of underutilized grasses

Many of underutilized grasses have been recognized to be resilient to numerous abiotic stresses including heat stress, drought, water logging, salinity, heavy metal toxicity etc. These also offer multiple uses including food, renewable fuel, feed, fiber and contributions to ecosystem services. A variety of underutilized grasses including reed canary grass (*Phalaris arundinacea* L.), miscanthus (Miscanthus × giganteus Greef et Deuter), giant reed (*Arundo donax* L.), switchgrass (*Panicum virgatum* L.) etc. have the potential to serve as an excellent raw material source for modern biorefineries for producing numerous high-added value products including nutrient supplements, biopharmaceuticals, biopolymers, biomaterials for mulching, building infrastructure, phonic insulating, biodegradable products for utilization in animal bedding and gardening, energy carriers including advanced biofuels, many by-products including green chemistry products and soil organic fertilizers, along with a bunch of ecosystem services such as soil erosion and degradation protection, C-sequestration, restoration and preservation degraded and contaminated soils. It has indicated that underutilized grasses have the potential to thrive well under variable agro-environmental conditions including degraded and marginal lands without being in competition with food crops. Besides higher environmental sustainability, bio-energy potential has also been recognized as a plus point of underutilized perennial grasses which are established once and provide harvest on a yearly basis over a period of 10–25 years resulting in greenhouse gas balances. The lignocellulosic structure of grass cell walls constitutes one of the critical sustainability characteristics which impart natural resistance against various pests and diseases [13].

Additionally, grasses tend to have higher resource-use efficiency for having C-4 photosynthetic pathway which is characterized by substantially higher solar radiation capture and moisture utilization, along with being lesser nutrient demanding and have potential to conserve nutrients in underground roots during harsh climatic conditions like chilling temperatures during winters. Furthermore, many underutilized grasses by virtue of their vigorous biomass production add crop residues to the soil due to senescence, and thus provide natural mulch for controlling weeds

Grasses	Technical name	Geographical presence	Perspective uses
Feather lovegrass	*Eragrostis amabilis*	Indo-Pak subcontinent, China, South Africa	Alternate forage and preserved feed (hay and silage) for ruminants
Job's tears	*Coix lacryma-jobi*	Philippines	Food products (porridge, coffee, wine, biscuits and variants of bread) and medicinal uses (wounds, blisters, and urinary tract infections)
Bermuda grass	*Cynodon dactylon*	Indo-Pak subcontinent, China, South America	Alternate forage and preserved feed for ruminants
Japanese sweet flag	*Acorus gramineus*	United States of America	Low-cost and environment friendly ornamental grass
Pycreus grass	*Pycreus flavidus*	Indo-Pak subcontinent, China, South Africa, Afghanistan, Iran, Iraq, Israel, Lebanon, Syria, Turkey	Alternate forage and preserved feed for ruminants
Hairy crabgrass	*Digitaria sanguinalis*	Indo-Pak subcontinent, China, South America	Forage and preserved feed (hey and silage) for ruminants
Miscanthus	*Miscanthus sp*	Mediterranean countries	Bioenergy production
Signalgrass	*Brachiaria racemosa*	Indo-Pak subcontinent, China, South Africa, Australia, Southern Europe	Forage and preserved feed for ruminants
Switchgrass,	*Panicun virgatum*	Mediterranean countries	Biomass crops for biofuel production
Wild sugarcane or Kans grass	*Saccharum spontaneum*	Indo-Pak subcontinent, Nepal, Bhutan, Panama	Fencing of houses, vegetable gardens, Thatching of houses or huts roofs
Giant reed	*Arundo donax*	Mediterranean countries	Biomass crops for biofuel production, phytoremediation of soil
Reed canary grass	*Phalaris arundinacea*	Mediterranean countries	Biomass crops for biofuel production
Lemon grass	*Cymbopogon citratus*	Philippines, Indonesia, Srilanka, Indo-Pak subcontinent, United Kingdom, Madagascar, Central America	Brewed into tea, use as herb in cooking for aroma, essential oils extraction and medicinal uses (antipyretic, antibacterial, and antifungal agent)
Chinese silvergrass	*Miscanthus sinensis*	United States, South American countries	Low-cost and environment friendly ornamental grass
Blood grass or cogon grass	*Imperata cylindrica*	United States, South American countries	Low-cost and environment friendly ornamental grass
Eastern gamagrass	*Tripsacum dactyloides*	North America	Bioenergy production
Prairie cordgrass	*Spartina pectinata*	North America	Bioenergy production
Big bluestem	Andropogon gerardii Vitman	North America	Biofuel production
Pink muhly grass	*Muhlenbergia capillaris*	United States of America, South American countries	Ornamental grass with esthetic values

Grasses	Technical name	Geographical presence	Perspective uses
Sand bluestem	Andropogon hallii Hack.	North America	Bioenergy production
Little bluestem	[*Schizachyrium scoparium* (Michx.) Nash]	North America	Biofuel production

Table 1.
Different underutilized grasses, their geographical presence and perspectives uses under varying farming systems and socio-economic perspectives.

and release of nutrients from residues after decomposition. Overall, cultivation of high yielding underutilized grasses can potentially multiply land-use-efficiency and higher productivity per unit of area. In addition, there is a great potential to still improve their performances. However, many of underutilized grasses are either undomesticated or are at earlier development stages, while in-depth studies are needed to develop their production technology package. One of the limitations of traditional breeding is the exceptionally lengthy process which might extend for over 15 years involving collection of germplasm, selection of parental lines, selective crossing to achieve desired traits and allowing evaluation cycles for random genetic mutations.

In addition to bio-energy applications of underutilized grasses, there are diversified uses of perennial grasses such as pulping as well as bleaching potential of giant reed for papermaking due to having moderate strength properties along with bleachability characteristics. In addition, miscanthus which is an underutilized grass has proved its potential and feasibility for producing various types of panel boards, building blocks of various infrastructures and medium-density fibreboard having comparable characteristics as that of wood chips. A significant equity between miscanthus and crops straw for animal bedding preparation could be achieved as far as cow comfort in the barnyard is concerned. However, the superiority of this perennial grass over straw has been established owing to higher biomass production potential compared to many cereals such as wheat, rice etc. Moreover, lignocellulosic biomass yielded by underutilized grasses might be processed into diversified products; however, lack of market development has so far hampered wider-scale implementation of the lignocellulosic biorefinery. It may be noted that lignocellulosic biomass currently fetches around 65 € per dry ton [12–14].

Thus, it becomes evident that productivity potential and the ability to generate comparable revenues of underutilized grasses would be the key drivers in farmer's perspectives. It also follows that comprehensive real time data pertaining to yield would be critical in order to provide accurate and reliable information to researchers, growers and entrepreneurs. Moreover, underutilized grasses future will be determined on development of consistent, feasible, farmer's friendly and affordable economic plans encompassing economically profitable plantation size and tailor-designed low-tech and easily accessible processing plants for producing market capturing products. At farm scale, advanced research for developing agronomic packages, designing breeding programs, building post-harvest logistics and bioconversion facilities are fundamental aspects that need thorough attention of researchers and governments for harnessing the potential of underutilized grasses. These will follow production of climate resilient genotypes of underutilized grasses having the potential to thrive well in a wider range of agro-environmental conditions on marginal lands without coming into competition with food crops (see **Table 1**).

5. Need of underutilized grasses promotion

Many underutilized grasses if promoted appropriately hold the potential to gain local, regional, national importance in terms of generating economic activity. For promoting underutilized grasses, securing resource base in developing countries is vital for maintaining the safety net comprising of diversified products and food stuffs and thus contributing to ensuring food security strives. Another justification for promoting underutilized grasses is to ensure diversification of agricultural systems and to offer support to fragile social groups having lesser affordability to rely solely on staple commodities [1, 14]. In addition, underutilized grasses cultivation on marginal, degraded and fellow lands can serve as a poverty alleviation strategy by empowering marginal sections of farming community. These also hold bright perspectives by allowing rural communities to adopt resources-based development instead of commodity focussing development. Along with poor segments of farming community, underutilized grasses can offer additional benefits to wider strata of communities through provision of balanced diets, diversified source of income to growers and marketing agents, sustainable preservation of agro-ecosystems and putting into use large swathes of marginal lands without disturbing the cultural identity.

6. Challenges posed to production of underutilized grasses

In designing research works and developing promotion programs for underutilized grasses, researchers and policy makers need to be prepared for coping multitude of problems and hindrances.

1. Lack of agricultural market globalization for novel and new agricultural products and overemphasis on a specific set of trade preferences for a certain number of cash or food crops is one of the key challenges in boosting the demand and utilization of products prepared from underutilized grasses. Lack of globalized market for novel products prepared from underutilized grasses serves as discouraging factors to researchers, growers, funding agencies and policy makers despite the fact that product development follows the market demand and keep on waiting for favorable market factors is bound to serve no purpose.

2. Urbanization and the associated promotion of large enterprises which have replaced small-scale commercial and economic activities can also make the promotion of underutilized grass products a daunting task by offering severe competition.

3. The homogenization of local cultures owing to the intensive interaction of diversified cultures by virtue of improved communication have further promoted specialization in favor of established crops and thus narrowing down the scope for entry of products developed from underutilized grasses having a comparative competitive disadvantage in terms of less market demand and little share in global trade.

4. Species selection of underutilized grasses constitute another big challenge as the right species selection from a broad genetic pool of potential candidates can ensure appropriate use of limited resources. The availability of incomplete and poor-quality information regarding localized grass species have further

multiplied the complexity on the selection process. It may be suggested that direct involvement of end users might be considered for a successful selection of underutilized grasses species.

5. Another daunting challenge is securing the necessary resource base for developing ex situ and in situ approaches intended for appraising the genetic diversity and running genetic programs in order to bring desired traits enabling the underutilized grasses to survive under changing climate scenarios.

7. Role of agronomy in underutilized grasses promotion

Underutilized grasses hold bright perspectives in improving the food and nutritional security of a rapidly increasing population; enhance the nutritional balance and impart sustainability to modern profit-oriented farming systems. These may also serve as grower-friendly poverty alleviation strategy by generating additional income and that too with the utilization of meager resources. Agronomy is a branch of agriculture which deals with sustainable production of food, feed, fuel and fiber crops by putting into practice biologically viable and economically attractive approaches encompassing persistently evolving production technology package and agricultural soil management, restoration and preservation. Agronomists hold critical role in boosting underutilized grasses cultivation on large scale by developing environmental friendly technology packages. Following are some of the vital roles that Agronomy and Agronomists can perform to make cultivation of underutilized grasses economically viable under changing climatic scenarios.

7.1 Bridging awareness and knowledge gaps

There exists serious research and knowledge gaps regarding the growth habits and input requirements of underutilized grasses, which have served as major constraints to the strives for promoting cultivation and creating demand of products developed from underutilized grasses. For time being, efforts are needed for raising awareness among stakeholders and encouraging them to execute research on underutilized grasses in order to redress their neglect status. Another aspect that needs thorough attention is to conduct a detailed analysis on the evolving status of species from underused grass to a well utilized crop. It becomes even more pertinent to develop criteria of a peak promotion stage at which a specific grass will cease to be underutilized. Agronomists need to shoulder the responsibility for promoting grasses that are not only biologically viable but also economically attractive to local farmers, keeping in view their technological level and size of landholdings. In addition, such a promotion package must also encompass boosting local diversity of flora and impart sustainability to production systems without compromising established farming systems and promoting cultivation of underutilized grasses on marginal lands and degraded soils that cannot support other crops of economic significance.

7.2 Access and multiplication of seeds

Planting material constitutes one of the most critical factors in determining the success of any crop and the same is the case of underutilized grasses. Agronomists would be required to strive for securing the genetic resources for establishing a diversified genetic pool of different underutilized grasses, and thereafter, intensive research might be conducted for screening out high yielding, climate resilient, stress tolerant and resource efficient genotypes. It will follow mass seed production through well planned and target-oriented seed production programs in order to improve the access of growers to quality seed at an affordable cost. Agronomists and plant breeders are required to work in liaison to run commercially feasible seed production programs.

7.3 Conservation through use

It must be conceded that resources have always been insufficient and scarce for conserving underutilized crops and grasses at large scale. Thus, this situation and desire for process sustainability requisite integration of conservation and utilization simultaneously and this concept is famously known as conservation through use. However, collecting information pertaining to distribution patterns, utilization preferences, and evaluation of existing traditional and localized knowledge on underutilized grasses is prerequisite as this information can serve as foundation of future research programs for improving access of growers to planting materials.

7.4 Localized agronomy

Since the cultivation and utilization of most of the underutilized grasses are primarily localized, thus agronomic packages for the cultivation of underutilized grasses must be developed keeping in view locally available farming resources, growers' technical know-how level, farm mechanization status, community needs and market scenarios. Local mechanisms that support the deployment of useful diversity will need to be strengthened. There is a dire need to develop integrated chains and networks that cohesively link Agronomists to farmers and end-users of products developed from underutilized grasses. Moreover, Agronomists need to work in liaison with agri-economists for assessing potential revenue generation from the cultivation of underutilized grasses and product development.

7.5 Future agronomic strives needed

The underutilized grasses have the potential to provide livelihoods to thousands of farmers globally provided access to planting material and quality is ensured along with investing in infrastructure development for product development through creating the market demand of products (food, feed, fuel, fiber, medicinal, spices, aromatic, beverages, esthetic etc.) developed from underutilized grasses. The underutilized grasses are threatened biological assets having the potential to contribute significantly in poverty alleviation strives, while Agronomists hold the key to unlocking this unutilized treasure. Intensive and systematic liaison among Agronomists and policy makers for developing cost-effective production technology package under specified set of agro-environmental condition and soil productivity status through appropriate resource allocation is the need of the time. Agronomists by developing biologically viable farming approaches can convert underutilized grasses into high value commodities for meeting community's real needs. Agronomists need to develop criteria regarding (a) the minimum technical know-how of growers required for successful cultivation of underutilized grass species, (b) compiling information that is easy to understand and repeat in different regions of the world for seed multiplication methods and real-time assessment of grasses regeneration capacity and (c) fundamental knowledge compilation on different types of insect-pests, diseases, and site-specific cultivation related hindrances. In addition, fundamental changes are needed in reporting the agricultural statistics at regional, national as well as international levels. It might be suggested that agricultural statistics year book that is compiled and published by the Food and Agriculture Organization (FAO) must be broadened in scope by adding underutilized crops and grasses. Besides compilation, wider and easy access to this information must be made available to researchers, extension workers, industry and other stakeholders. At local and regional levels, site-specific studies to optimize production technology package of underutilized crops must be supported for ensuring their publication in order to make results available to wider audience.

8. Crop protection role in promoting underutilized grasses

Crop protection constitutes one of the most vital branches of agricultural science which keeps on devising biologically viable ways and cost-effective means for controlling various types of diseases, insect-pests in order to prevent significant damage by keeping harmful organisms below threshold levels. To prevent a severe disease outbreak comprises maintaining a healthy and vigorously growing crop. Each individual plant in the field requires optimum water and fertilizer quantity, as well as an aerated, well-drained soil but lacking any of these factors, the crop may become stressed ultimately more susceptible to disease. A study revealed that microbial diseases are responsible for the ultimate crop losses up to 16%, out of these 16% microbial losses almost 70–80% were due to fungal pathogens. It is estimated that more than 100,000 plant diseases can be caused by 8000 reported fungal species. As far as the underutilized grasses needs to be maintained by characterization and research on its agronomic factors, still there is a dire need to explore the pathogens causing mild to severe diseases ultimately suffering a huge loss in its production and quality traits. A few of the major crops may responsible for nutrition as well as food security that ultimately leads to keep the agriculture system vulnerable to various biotic and abiotic stresses due to the lack of genetic diversity in these crops. As far as diseases are concerned, there may be fungal and viral diseases that may be challenging to adopt in the underutilized grasses [15–18].

Besides numerous diseases, a few need more attention as to be more severe in the grasses which must be investigated to find out biologically viable solution for keeping these below the threshold level. Rust caused by the species of genus Puccinia and is obligate plant pathogen. This genus contains more than 4000 species based on their hosts. Considering lemongrass as an example of underutilized grasses the rust caused by *Puccinia nakanishikii* Dietel more sever in warmer and more humid areas. It produces light brown pustules on both the lower and upper surfaces of leaves. The spores dispersal through wind may spread the disease on larger scale. Unfortunately, there is still the lacking research on the management strategies of this disease on lemongrass and is a dire need to address this issue to overcome the pathogen potential. Furthermore, *Helminthosporium cymbopogi* another fungal pathogen causing a sever disease of grasses including lemongrass known as leaf spot. Similarly leaves curling and browning caused by brown tip disease is due to the low water content in the leaves. Foliage blight is another fungal disease caused by *Curvularia andropogonis* (Zimm.) infecting mostly grasses led to the considerable yield losses. The common management practice to control these fungal diseases is application of 1% Bordeaux mixture or 0.3% Zineb three times with an interval of fifteen days. Similarly, 0.2–0.3% Mancozeb can be an alternative fungicide application thrice in the season with 15 days interval [19–23].

Similarly blast is another important fungal disease on grasses especially on millet caused by *Pyricularia grisea* lead to sever grain losses 56–80% while upto 35% losses were reported in 1000-grain mass. Millet is vulnerable to this pathogen from seedling till its grain formation. Commonly the symptoms are spindle shaped lesions of different sized, generally the spots appear initially with yellowish margins and gray centers. The lesions later on turned to whitish gray and also olive gray growth of fungus may appear on the lesions. Seed treatments with Tricyclazole may be effective to overcome the primary seed born inoculum. Later on, fungicide application on ear appearance and after 10 days interval should give better results. It has been reported that biological control agent 0.6% *Pseudomonaas fluorescens* used as seed treatment following two later spays of the same bio-agent may constitute a good alternative to chemical fungicides for underutilized grasses [24–26].

Among nematode disease cereal cyst nematodes among one of the oldest genus named Heterodera are the more important that may infect small cereal grain crops like oat, barley, wheat, rye, and triticale. Cereal cyst nematodes complex widely distributed on family Poaceae includes several species. Among these species oldest reported specie was *Heterodera avenae* followed by *H. latipons*, then H. *hordecalis* in North Europe, furthermore H. filipjevi in eastern Europe, up till now 11 species of genus *Heterodera* has been reported. Among these 11 species three of them i.e. *H. avenae*, *H. latipons*, and *H. filipjevi* considered economically important on cereals globally [25–28].

Generally, the best management practice to normalize the effect of cereal cyst nematode may include crop rotation with a non-host crop. The eggs of cyst may become dormant inside the cyst for many years but have a very narrow host range, therefore rotation led to the best cultural practice. Furthermore, clean fallows, sanitation of fields, weed control, sowing time to escape egg hatching and trap cropping should be effective. Use of resistant varieties and chemical nematicides directly minimize the population density of nematode. Studies revealed that the use of nematophagous fungi should be an alternative of chemical control as to target the cyst nematodes with the use of these biological control agents.

9. Food sciences contribution in boosting underutilized grasses products

Food science belongs to basic as well as applied sciences of food and its scope significantly overlaps with agricultural science along with nutritional science leading through different vital scientific aspects pertaining to food processing and food safety along with persistent development of economically feasible technologies for food processing. Regarding underutilized grasses, food sciences can potentially play a vital role through product development and creating market demand for food products developed from underutilized grasses like lemon grass. There is an increasing pressure on agriculture to produce greater yields of feed, food and biofuel from limited land resources for the estimated population of nine billion people on the globe by 2050 [1–3]. So it is proposed that production from agricultural sources has to be increased to manage an estimated 40% increase in the world's population. About 90% of this progress is likely to result from improved cropping and high crop yields, while the remaining has to be produced from land resources presently not utilized for farming. The diversification of crops from the poaceae family having nutritional value can also cope with the problem of food insecurity. The diversification into other grass crops could lead to sustainable agriculture by enhancing economic, ecological, nutritional and social conditions.

9.1 Underutilized grasses for sustainable food production and nutritional security

In comparison to the staple grasses, neglected or undervalued grasses are of immense importance in the food industry for developing valuable products. Food

Science and technology has a wide range of applications for utilizing different undervalued grasses for the production of edible sugars and glucose from non-used rice and wheat residues which are usually wasted or burnt, and has been recently introduced for the successful production of food grade sugars. Similarly, Green Grass juice from underutilized barley and wheat is another essentially therapeutic food product with functional food ingredients. Green juice from grasses contains chlorophyll which is considered as green blood as it is a substitute of hemoglobin. So, the utilization of these grasses for juice production in the juice industry can achieve a wide range of objectives, including maintaining consumer's health. The active ingredients in these juices also hold functional properties of immense importance for the juice processing industry. However, their contributions should be studied in order to enhance the precision. Cereal grass juices must be encouraged as a functional beverage in diet-based therapies against different lifestyle-related disorders [29–31].

Another wonderful candidate from Poaceae family is lemon grass. Lemongrass is primarily cultivated and grown for its essential oil (EO) that has multiple medicinal (anticancer, analgesic and antimicrobial) and cosmetic uses. It is also utilized in the form of herbal tea (green tea) as it contains a variety of vitamins and minerals which are essential for health. Lemongrass derivatives in aqueous or dried extract form can be used for the preparation of acceptable mixed beverages. This valuable product could be developed to improve the antioxidant activity, nutritional aspects, and health benefits. Usually, grains from grasses are utilized but grasses are ignored as a waste. Barley grass powder has a huge potential of utilization as a functional food ingredient in food preparations. Barley grass is rich in vitamins and minerals and can be developed as a powdered supplement to treat many chronic diseases. Also, food industries can utilize the barley powder for fortification purposes. Infect some effective strategies of food scientists are required that can guide futuristic research on production of functional foods from barley grass for prevention and treatment of chronic diseases [32].

Recently, Denmark's National Food Institute contributed towards the novel application of grass protein as a food for human consumption. As it will be a cheaper and valuable source of protein to cope with the issues of food insecurity and alternatively protein deficiency malnutrition around the globe [33]. Interestingly, grass protein powder is a profitable and sustainable concept for serving humanity on the earth. Researchers claimed that grass protein has similar amino acid profile to that of egg, soya and whey proteins. For grass proteins ryegrass is an ideal candidate as it contains the right amino acids composition that can be turned out as a good protein source for human consumption. It is of prime interest that protein powder from rye grass can be utilized in a wide range of food products. As a novel food item grass protein powder must be approved by the European Food Safety Authority to ensure the powder is safe for human consumption. Researchers and food technologists are ambitious to develop grass protein as a food ingredient as it will be of great contribution towards an economical, approachable and sustainable solution to solve Food insecurity issues.

10. Plant breeding and molecular approaches for underutilized grasses promotion

Plant breeding is a set of scientifically driven procedures and techniques for developing new genotypes through process called crop improvement, cultivar development and seed improvement. It assists to create multi-generations of genetically diverse populations generally through human triggered selection for creating the adapted plants having new combinations of desirable traits. There is an urgent need of developing the grass species having potential to yield higher under rapidly changing climate scenarios. This can be achieved by imparting traits of tolerance against abiotic and biotic stresses in order to fulfill the rising demands of food for rapidly growing populations. In the mid era of twentieth century, conventional breeding methods of plants had resulted in the historical green revolution since very high yielding crop varieties were produced by breeders. However, now under the scenario of climate change, conventional methods of breeding plant species are not sufficient. Molecular tools and techniques have evolved for developing plant-species with enhanced nutritional value through direct transfer of desirable genes controlling the demanded traits. Genetically engineered or modified crops, conventionally named the genetically-modified-crops (GMOs) can effectively supplement the conventional methods for producing improved quality plants for food and feed. Crop-species can be developed by genetic engineering for enhanced yield, nutritional qualities as well as the enhanced resistance to different environmental stresses. Breeding strategies for improved forage species is different from major crops since it requires a long-duration and demands the integrated use of the other disciplines such as; genetics, breeding, biotechnology, agronomy, entomology, physiology, pathology and animal-nutrition [34].

Breeding programs for underutilized grass species require complete knowledge of species-genetic-relationship, chromosomal composition, polyploidy and the, degree of existing gene-recombination or genetic variation for further selection and hybridization. Hence, the overall strategy differs among the species. However, a remarkable progress in the areas of modern molecular gene engineering tools has opened new horizons. Molecular approaches using biotechnological tools to produce improved forage crop varieties were started in the late-eighties. Such biotechnological tools include: Molecular techniques to observe the genetic composition, foreign or distant-gene insertion directly into the targeted plant-genome, and micro propagation from single cells in vitro. Various other such techniques such as embryo rescue, haploid plant production and creation of new variations aid in different steps involved conventional breeding methods consequently minimizing time required for conventional breeding methods. Additionally, the plants bred through such techniques do not conflict with the interests of the individuals who oppose the genetically -modified -organisms. For production of hybrids of Lolium-Festuca, the embryo-rescue technique has been exploited efficiently. There are several classic techniques of molecular breeding viz.; restriction-fragment-length polymorphism (RFLP), random amplified polymorphic DNA (RAPD), amplified- fragment-length polymorphism (AFLP), and isozymes which are frequently exploited for characterization of germplasm, quality trait loci (QTL) identification, detection of hybrids, cultivar identification, gene tagging, and genetic mapping. The molecular characterization of the genetic structure of forage crops as well as weeds is equally important. Since, if the gene identified from one plant species or living-organism contains the similarity in its sequence offers ease in its transfer into the target species through gene transformation [35].

Although, characterization of available germplasm is crucial particularly under the changing climates scenario, the gene-tagging and genetic-mapping in forage species is much lagging behind. For traits which are under the control of a single gene, gene tagging is essential, but in the case of forages most of the desirable agronomic traits are under the control of many genes and are thus very difficult to tag. Gene identification for the genes controlling apomixes in grass-breeding is a key to produce hybrid seed of underutilized grass species. Cloning and functional identification of these genes can be patented by breeder and can also be used for fixing heterosis in various species and offers time saving for hybrid seed production each year. Famous example is the Napier x Bajra hybrid, which was produced by the cross between *Pennisetum glaucum* and *Pennisetum purpureum*.

Recent progresses in the areas of genomics complemented with high-throughput and precision phenotyping facilitate the identification of genes controlling economic agronomic traits. The detection of these genes can be combined with genome editing techniques for the speedy development of climate change resilient plant species. Currently, genome editing is applied in major food crops and this technique has the potential for rapid improvement of underutilized crop plants, specially, targeting the current and future challenges of climate change. The success of genomics in improving a given plant species is also influenced by the nature of the trait under study. For example, traits intensely affected by the environment and genotypic and the environmental interaction are more challenging to study and modify [36]. Another approach could be intercropping underutilized grasses with staple cereals and legumes as this approach has the potential to boost soil fertility, total yield, and economic turnouts along with numerous other ecological benefits such as improvement in soil microbial population [37–40].

Transgenic technology allows the transfer of foreign genes from unrelated species and thus offers enormous scope to improve underutilized grass species. The development of more detailed gene maps of different species, using genomics and allied molecular tools will help in the identification of genes or gene sequences that might be associated with responses to changing climate stress. Although the biosafety and health hazards linked with GM crops have been questioned, a number of crop species have already been genetically-engineered and carefully tested and possess no obvious risk. Integrated use of modern biotechnology, with conventional agricultural in a sustainable way, can lead to achieving the ultimate goal of achieving food security for current and future populations. Transgenic approaches have been employed to improve these species in the following aspects: significant improvement of dry matter digestibility in the case of tall fescue, alfalfa, and perennial ryegrass. By efficient integration of novel germplasm into practical breeding programs, transgenic cultivars offer the potential to play a potential role in fulfilling the growing demand for animal products as well as renewable fuels in the coming years.

11. Pertinence of participatory approach and agri-sciences integration

A participatory approach integrating different disciplines including Agronomy, crop protection, plant breeding molecular genetics and food sciences to promote the cultivation and market demand of products developed from underutilized grasses is the need of time. It becomes even more important as studies on underutilized grasses have remained neglected historically and constitute one of the biggest challenges in crop genetic resource history. The destiny changing phenomenon of the green revolution holds witness to the fact that inter-disciplinary and trans-disciplinary approaches integrated in a coherent way to boost underutilized grass production is one of the most feasible, doable and viable options. It must be recognized that underutilized grass species will never command the same prime undertaking as a major crops which requisites a different but integrated approach for their viable promotion. Such an approach must link all stakeholders and research activities pertaining to local grass agro-botanical and pathological information collection, research trials, product development, nutritional assessment of developed products for safety and taste, product utilization policy and marketing as well as commercialisation plans. A chain of researchers from Agronomy, crop protection, plant breeding and food sciences can conduct interconnected research for boosting cultivation of underutilized grasses and develop products keeping in view the needs and demands of local, regional, national and international markets.

There is a very critical role of international organizations such as FAO for the sharing of findings acquired in region with equal benefits to other regions in terms of grasses cultivation and product development. The participatory approaches formulated collectively through larger brainstorming among stakeholders and implemented under localized conditions occupies the strategic position for making the best utilization of existing resources and promoting synergism across different regions. Underutilized grasses also constitute a class of grasses that are ignored socially and therefore, generalized masses and farmers are bound to attract towards multi-disciplinary research teams instead of working in isolation. The inter-disciplinary and multi-disciplinary researchers put a halt to the persistent decline in genetic erosion of grasses. Even extension workers can perform strategic role by collecting information regarding underutilized grasses from farmers of far-flung areas and thereafter Agronomists and Food Technologist can work cohesively to reveal the true potential of underutilized grasses through the production of quality products having rich perspectives in localized and regional markets.

Additionally, Agronomist need to work in loop with crop protection research-ers to analyze the constraint factors related to insect-pest and diseases incidence, leading to the development of a technology package enabling grasses to cope with biotic and abiotic stresses effectively under a changing climate. The participatory approach involving Agronomists with Breeders may contribute to enhancing the see and germplasm selection, production, multiplication, supply, processing, product development and commercialisation. Furthermore, inclusive strategies hold the perspectives to develop rapid marketing demand for products from underutilized grasses through intensive cooperation with the private sector. The participatory approaches must attempt to explore options to grasses conservation and use simultaneously in order to secure a resource base for boosting underutilized grass cultivation and production. The approaches may differ, depending on whether the crop is seed propagated or clonally propagated, annual or perennial, outbreeding or self-pollinated. It is worth mentioning that a participatory approach must encom-pass information on the smallest size of ex situ collection that may ensure genetic diversity along with the ways and techniques to economically maintain the genetic diversity. Moreover, it is also vital to determine the extent of diversity that must be included in the production systems along with developing monitoring criteria in order to make the successful cultivation of underutilized grasses on a wider scale.

Besides agronomic packages, technologies entailing molecular genetics and GIS might play their role in developing the conservation techniques and utiliza-tion strategies for underutilized crops. As implied in the case of inter-disciplinary and trans-disciplinary approaches, it is also needed to initiate sustainable linkages among researchers, research and development organizations, farmers and consum-ers. It is always unlikely that researchers belonging to a specific discipline have all the expertise, while any single organization can also ill-afford to support research work on a large scale for boosting underutilized grass production and product development. Ultimately, it must be recognized that underutilized grasses present unique a set of problems and potential opportunities under varying socio-economic conditions, and thus participatory approaches can improve conservation and utilization of underutilized grasses under changing climate scenarios.

12. Conclusions

The commercial-oriented farming systems encompassing the cultivation of major grasses have caused a serious decline in the intra and interspecific diversity of crops. In addition, the decline of grasses biodiversity has led to higher vulnerability among growers and end-users while the changing climate has made it mandatory to promote underutilized grasses (Feather lovegrass, job's tears, bermuda grass, Japanese sweet flag, pycreus grass, hairy crabgrass, signalgrass, switchgrass, miscanthus, giant reed, reed canary grass, lemon grass, Chinese silvergrass, big bluestem, wild sugarcane etc.) diversity in order to ensure food security and economic viability of modern farming systems. The panacea lies in a participatory approach entailing integration of agronomic practices with crop protection, food sciences and plant breeding in order to develop sustainable technology packages for ensuring economic production of food, beverage and medicinal products from underutilized grasses. Moreover, creating market demand for novel products of underutilized grasses coupled with sustainable supplies of raw material along with processing, packaging and branding facilities hold key in booting cultivation and utilization of underutilized grasses under changing climate. Last but not least, United Nation's envisaged sustainable goals of zero hunger and poverty alleviation might also be addressed by boosting cultivation and utilization of underutilized grasses.

Author details

Muhammad Aamir Iqbal[1*], Sadaf Khalid[1], Raees Ahmed[2],
Muhammad Zubair Khan[3], Nagina Rafique[4], Raina Ijaz[5], Saira Ishaq[4],
Muhammad Jamil[1], Aqeel Ahmad[1], Amjad Shahzad Gondal[6], Muhammad Imran[7],
Junaid Rahim[7] and Umar Ayaz Aslam Sheikh[7]

1 Faculty of Agriculture, Department of Agronomy, University of Poonch
Rawalakot, Pakistan

2 Faculty of Agriculture, Department of Plant Pathology, University of Poonch
Rawalakot, Pakistan

3 Faculty of Agriculture, Department of Plant Breeding and Molecular Genetics,
University of Poonch Rawalakot, Pakistan

4 Faculty of Agriculture, Department of Food Science and Technology, University
of Poonch Rawalakot, Pakistan

5 Faculty of Agriculture, Department of Horticulture, University of Poonch
Rawalakot, Pakistan

6 Department of Plant Pathology, Bahauddin Zakariya University, Multan, Pakistan

7 Faculty of Agriculture, Department of Entomology, University of Poonch
Rawalakot, Pakistan

*Address all correspondence to: aamir1801@yahoo.com

IntechOpen

References

[1] Kizilgeci F, Yildirim M, Islam MS, Ratnasekera D, Iqbal MA, Sabagh AE. Normalized difference vegetation index and chlorophyll content for precision nitrogen management in durum wheat cultivars under semi-arid conditions. Sustainability. 2021;**13**:3725

[2] Abbas RN, Arshad MA, Iqbal A, Iqbal MA, Imran M, Raza A, et al. Weeds spectrum, productivity and land-use efficiency in maize-gram intercropping systems under semi-arid environment. Agronomy. 2021;**11**:1615

[3] Haque MM, Datta J, Ahmed T, Ehsanullah M, Karim MN, Akter MS, et al. Organic amendments boost soil fertility and rice productivity and reduce methane emissions from paddy fields under sub-tropical conditions. Sustainability. 2021;**13**:3103

[4] Chowdhury MK, Hasan MA, Bahadur MM, Islam MR, Hakim MA, Iqbal MA, et al. Evaluation of drought tolerance of some wheat (Triticum aestivum L.) genotypes through phenology, growth, and physiological indices. Agronomy. 2021;**11**:1792

[5] Iqbal MA, Rahim J, Naeem W, Hassan S, Khattab Y, Sabagh A. Rainfed winter wheat (Triticum aestivum L.) cultivars respond differently to integrated fertilization in Pakistan. Fresenius Environmental Bulletin. 2021;**30**(4):3115-3121

[6] Alghawry A, Yazar A, Unlu M, Çolak YB, Iqbal MA, Barutcular C, et al. Irrigation rationalization boosts wheat (Triticum aestivum L.) yield and reduces rust incidence under arid conditions. BioMed Research International. 2021;**2021**. Available from: https://ops.hindawi.com/view.manuscript/bmri/5535399/1/

[7] Hakim AR, Juraimi AS, Rezaul Karim SM, Khan MSI, Islam MS, Choudhury MK, et al. Effectiveness of herbicides to control rice weeds in diverse saline environments. Sustainability. 2021;**13**:2053

[8] Iqbal A, Iqbal MA, Awad MF, Nasir M, Sabagh A, Siddiqui MH. Spatial arrangements and seeding rates influence biomass productivity, nutritional value and economic viability of maize (Zea mays L.). Pakistan Journal of Botany. 2021;**53**(3):967-973

[9] Alam MA, Skalicky M, Kabir MR, Hossain MM, Hakim MA, Mandal MSN, et al. Phenotypic and molecular assessment of wheat genotypes tolerant to leaf blight, rust and blast diseases. Phyton, International Journal of Experimental Botany. 2021;**90**(4): 1301-1320

[10] Sorour S, Amer MM, El Hag D, Hasan EA, Awad M, Kizilgeci F, et al. Organic amendments and nano-micronutrients restore soil physico-chemical properties and boost wheat yield under saline environment. Fresenius Environmental Bulletin. 2021;**30**(9):10941-10950

[11] Singh K, Awasthi A, Sharma SK, et al. Biomass production from neglected and underutilized tall perennial grasses on marginal lands in India: A brief review. Energy Ecology and Environment. 2018;**3**:207-215

[12] Lewandowski I, Scurlock JM, Lindvall E, Christou M. The development and current status of perennial rhizomatous grasses as energy crops in the US and Europe. Biomass and Bioenergy. 2003;**25**:335-361

[13] Li C, Xiao B, Wang QH, Yao SH, Wu JY. Phytoremediation of Zn- and Cr-contaminated soil using two promising energy grasses. Water Air Soil Pollution. 2014;**225**:20-27

[14] Ahmad F, Hameed M, Ahmad MSA. In: Ozturk M, Hakeem K, Ashraf M, Ahmad M, editors. Exploring Potential of Minor/Underutilized Grasses for Remote Areas Facing Food Scarcity, Global Perspectives on Underutilized Crops. Cham: Springer; 2018

[15] Koike ST. Rust disease on lemongrass in California. Plant Disease. 1999;**83**:304

[16] Barua A, Bordoloi DN. Record of a new disease of lemongrass (Cymbopogon flexuosus Stapf.) caused by Curvularia verruciformis Agarwal and Sahni. Current Science. 1983;**52**:640-641

[17] Chutia M, Mahanta JJ, Sakia RC, Baruah AKS, Sarma TC. Influence of leaf blight disease on yield and its constituents of Java citronella and in vitro control of the pathogen using essential oils. World Journal of Agricultural Sciences. 2006;**2**(3): 319-321

[18] Sunil MB, Anilkumar TB. Effect of head blast on grain mass and grain color in finger millet. Sorghum Research Reports. 2004;**2**(3):163-170

[19] Patro TSSK, Rani C, Kumar GV. Pseudomonaas fluorescens, a potential bioagent for the management of blast in Eleusine coracana. Journal of Mycology and Plant Pathology. 2008;**38**(2): 298-300

[20] Nicol JM, Rivoal R, Taylor S, Zaharieva M. Global importance of cyst (Heterodera spp.) and lesion nematode (Pratylenchus spp.) on cereals: Distribution, yield loss, use of host resistance and integration of molecular tools. In: Cook R, Hunt DJ, editors. Nematology Monographs and Perspectives. Tenerife, Spain: Proceedings of the Fourth International Congress of Nematology; 2004. pp. 1-19

[21] Nicol JM, Rivoal R. Global knowledge and its application for the integrated control and management of nematodes on wheat. In: Ciancio A, Mukerji KG, editors. Integrated Management and Biocontrol of Vegetable and Grain Crops Nematodes. Dordrecht: Springer; 2008. pp. 243-287

[22] Franklin MT. Heterodera latipons n. sp., a cereal cyst nematode from the Mediterranean region. Nematologica. 1969;**15**:535-542

[23] Andersson S, Heterodera hordecalis n. sp. (Nematoda: Heteroderidae) a cyst nematode of cereals and grasses in southern Sweden. Nematologica. 1974;**20**:445-454

[24] Smiley RW, Nicol JM. Nematodes which challenge global wheat production. In: Carver BF, editor. Wheat Science and Trade. Ames: Wiley-Blackwell; 2009. pp. 277-284

[25] Dababat AA, Pariyar S, Nicol J, Duveiller E. Cereal Cyst Nematode: An Unnoticed Threat to Global Cereal Production. Ibadan: CGIAR SP-IPM Technical Innovation Brief; 2011. pp. 286-290

[26] Dawabah AAM, Al-Hazmi AS, Al-Yahya FA. Management of cereal cyst nematode (Heterodera avenae) in a large scale wheat production. In: Dababat AA, Muminjanov H, Smiley RW, editors. Nematodes of Small Grain Cereals: Current Status and Research FAO. Ankara: FAO; 2015. pp. 277-284

[27] Ashoub AH, Amara MT. Biocontrol activity of some bacterial genera against root-knot nematode, Meloidogyne incognita. Journal of American Science. 2010;**6**:321-328

[28] Godfray HCJ, Beddington JR, Crute IR, Haddad L, Lawrence D, Muir JF, et al. Food security: The

challenge of feeding 9 billion people. Science. 2010;**327**:812-818

[29] Bruinsma J. The resource outlook to 2050: By how much do land, water and crop yields need to increase by 2050? In: Proceedings of the Technical Meeting of Experts on How to Feed the World in 2050, Rome, Italy, 24-26 June 2009. Rome, Italy: Food and Agriculture Organization (FAO); 2009. pp. 1-33

[30] Aiza Q, Farhan S, Muhammad TN, Abdullah IH, Khan MA, Niaz B. Probing the storage stability and sensorial characteristics of wheat and barley grasses juice. Food Science Nature. 2019;**7**:554-562

[31] Yawen Z, Xiaoying P, Jiazhen Y, Juan D, Xiaomeng Y, Xia L, et al. Preventive and therapeutic role of functional ingredients of barley grass for chronic diseases in human beings. Oxidative Medicine and Cellular Longevity. 2018;**21**:117-121

[32] Dirlei DK, Prudencio SH. Blends of lemongrass derivatives and lime for the preparation of mixed beverages: Antioxidant, physicochemical, and sensory properties. Journal of Science of Food and Agriculture. 2019;**99**:1302-1310

[33] Datta A. Genetic engineering for improving quality and productivity of crops. Agriculture & Food Security. Agriculture & Food Security. 2013;**2**(1):15-23

[34] Pourkheirandish M, Golicz AA, Bhalla PL, Singh MB. Global role of crop genomics in the face of climate change. Frontiers in Plant Science. 2020;**11**:922

[35] Tomlinson I. Doubling food production to feed the 9 billion: A critical perspective on a key discourse of food security in the UK. Journal of Rural Studies. 2013;**29**:81-90

[36] Wang ZY, Brummer EC. Is genetic engineering ever going to take off in forage, turf and bioenergy crop breeding? Annals of Botany. 2012;**110**(6):1317-1325

[37] Iqbal MA, Iqbal A, Ahmad Z, Raza A, Rahim J, Imran M, et al. Cowpea [*Vigna unguiculata* (L.) Walp] herbage yield and nutritional quality in cowpea-sorghum mixed strip intercropping systems. Revista Mexicana De Ciencias Pecurias. 2021;**12**(2):402-418

[38] Iqbal MA, Hamid A, Hussain I, Siddiqui MH, Ahmad T, Khaliq A, et al. Competitive indices in cereal and legume mixtures in a south Asian environment. Agronomy Journal. 2019;**111**(1):242-249

[39] Iqbal MA, Hamid A, Ahmad A, Hussain I, Ali S, Ali A, et al. Forage sorghum-legumes intercropping: Effect on growth, yields, nutritional quality and economic returns. Bragantia. 2019;**78**(1):82-95

[40] Iqbal MA, Iqbal A, Abbas RN. Spatio-temporal reconciliation to lessen losses in yield and quality of forage soybean (*Glycine max* L.) in soybean-sorghum intercropping systems. Bragantia. 2018;**77**(2):283-291

Chapter 5

Miscanthus Grass as a Nutritional Fiber Source for Monogastric Animals

Renan Donadelli and Greg Aldrich

Abstract

While fiber is not an indispensable nutrient for monogastric animals, it has benefits such as promoting gastrointestinal motility and production of short chain fatty acids through fermentation. *Miscanthus x giganteus* is a hybrid grass used as an ornamental plant, biomass for energy production, construction material, and as a cellulose source for paper production. More recently Miscanthus grass (dried ground *Miscanthus x giganteus*) was evaluated for its fiber composition and as a fiber source for poultry (broiler chicks) and pets (dogs and cats). As a fiber source, this ingredient is mostly composed of insoluble fiber (78.6%) with an appreciable amount of lignin (13.0%). When added at moderate levels to broiler chick feed (3% inclusion) Miscanthus grass improved dietary energy utilization. However, when fed to dogs at a 10% inclusion Miscanthus grass decreased dry matter, organic matter, and gross energy digestibility, and increased dietary protein digestibility compared to dogs fed diets containing similar concentrations of beet pulp. Comparable results were reported for cats. In addition, when Miscanthus grass was fed to cats to aid in hairball management, it decreased the total hair weight per dry fecal weight. When considering the effects Miscanthus grass has on extruded pet foods, it behaves in a similar manner to cellulose, decreasing radial expansion, and increasing energy to compress the kibbles, likely because of changes in kibble structure. To date, Miscanthus grass has not been evaluated in human foods and supplements though it may have applications similar to those identified for pets.

Keywords: *Miscanthus x giganteus*, fiber nutrition, insoluble fiber, pet nutrition, human nutrition, pet food processing, fiber profile

1. Introduction

Fiber ingredients added to foods for humans and animals are typically co-products from the wood-pulp industry (cellulose), byproducts from cereal (*e.g.*, bran, psyllium), legume seed (pea fiber), and vegetable (*e.g.*, tomato pomace) processing. More deliberate fibers such as inulin, FOS, Chicory root extract and other prebiotics are also common to foods. Unintentional fibers such as those from gums and gelling agents (*e.g.*, carrageenan, guar gum) are used in processed foods. Seldom have the grasses or forages been considered for use in foods as a fiber additive for monogastric animals. This has been the domain of grazing animals and as supplemental feed during confinement for ruminants and hind-gut fermenters (*e.g.*, horses, rabbits), or used as bedding. However, forage grasses may be a viable alternative fiber source for monogastric animals under certain circumstances. Relative to the current options, the grasses would certainly qualify as less processed and could even be considered as a purpose grown, sustainable, low environmental impact ingredient in diets for man and animal. Miscanthus grass is one such novel grass that has been evaluated as a fiber source for broiler chickens, dogs, and cats [1–6]. Other authors have also evaluated this fiber for companion animal applications [7]. For purposes of this review, it is our goal to provide a comprehensive summary regarding the information available to date regarding the use of Miscanthus grass in monogastric animal food products with a nod to human nutrition. Additionally, an overview of existing knowledge regarding how this ingredient impacts food processing will be provided.

2. Materials and methods

The focus of this chapter was Miscanthus grass as a potential fiber source for monogastrics. A literature search was conducted with the aid of Google Scholar using the following search terms: Miscanthus grass, *Miscanthus giganteus*, dog, canine, cat, feline, chicken, poultry, pig, swine, food processing, particle size, and human. Literature published between 1950 and 2021 was selected as potential references to be used in this chapter. Other supporting literature related to the history, biology and agronomy of this crop was obtained from Google Scholar using search terms such as, but not limited to, *Miscanthus giganteus*, origin, cultivation, uses, production, NDF, ADF, ADL, TDF, insoluble fiber, soluble fiber, particle size, flowability. Other reference information available to the authors in the form of other texts, abstracts, and thesis were also considered.

3. *Miscanthus x giganteus* history and general characteristics

Miscanthus x giganteus is a hybrid plant created in Japan, likely by the combination of *M. sinensis* and *M. sacchariflorus* [8]. Presumably it was then brought to Denmark in the mid 1930's and spread throughout Europe and North America as a horticultural plant [8]. The hybrid is sterile; thus, its propagation is through viable rhizome plantings and spread (**Figure 1A**). In the past it was used as forage for animals and for thatching [11]. However, in recent years, it has been considered as a source of cellulose for fuel to produce heat and electricity [12] via ethanol production [9], as well as construction materials, and absorbents [13].

M. x giganteus is a C4 plant relying on the NADP-malic enzyme pathway [14]. This pathway allows for the continuous photosynthesis even at lower temperatures (8°C) [15]. This is an important characteristic that has allowed this plant to be successfully cultivated in colder climates, such as northern Europe and North America. Moreover, this plant efficiently uses nitrogen and water [16, 17] compared to other crops. Thus, while *M. x giganteus* has not been adapted to produce food, it does grow well in marginal soils which are not suitable for cultivation.

Some authors report that the plant once established can remain productive for 5 to 40 years [11, 18, 19] depending on the region in which it is cultivated and cropping pressure (**Figure 1B**). Thus, *M. x giganteus* is considered a perennial crop. In this state it grows quickly and reaches 2 m in height with a close canopy cover which reduces sun light penetration, limiting weed growth, thus eliminating the need for herbicide administration (**Figure 1B**). Although, weed control is necessary before this stage as the plant is getting established [20]. Nutrient use by *M. x giganteus* is very efficient as it translocates nitrogen, phosphorus, and potassium to

Figure 1.
Miscanthus x giganteus *rhizome (A; from Adams et al. [9]), growth stage approximately 2.5 m (B); dried (C; from Adams et al. [9]); baled (D; from Adams et al. [9]), stored bales (E), and ground (F; from Pontius et al. [10]) with a particle size of 134 ± 93 μm and a 5X magnification.*

the rhizomes at the end of the growing season when the aerial portion of the plant begins to senesce (**Figure 1C**) [16]. This senescence starts with a killing frost during fall [21]. Predation by insects is limited [22]. As a result, this plant has been primarily utilized for biomass production; although, there may be more value for this crop than has been identified to date.

In general, fiber rich ingredients have been gaining more attention. In part because obesity in the pet and human population is a substantial issue [23, 24] and fiber is one possible solution to decrease the energy density of food. It may also increase the volume of the digesta in the gastrointestinal tract, and the fermentation of fiber in the colon to short chain fatty acids like butyrate (a preferred fuel source for the colonocyte) may aid in the prevention of cancer and the reduction in intestinal inflammation [25]. Moreover, food fiber through bulking of digesta can help alleviate constipation [26]. Despite these health benefits, fiber-added foods are usually less preferred than "regular" foods [27, 28]. Part of the changes in the flavor and texture attributes of fibers could be related to the composition of various fiber sources. For example, lignin a phenylpropanoid component of some fiber ingredients is known to have a bitter taste [29]. An alteration to texture is likely an effect of the changes that fiber cause in the product during processing that changes the mouthfeel as the food is consumed [30]. However, acceptance of dietary fiber may be changing as consumers attribute more importance to the health benefits and their palates adjust to the flavor and texture profile of these more fibrous products.

Despite the health benefits and their popularity in some human and pet foods, adding fiber ingredients brings challenges to manufacturing. For example, in extruded expanded products (like breakfast cereals and dry extruded pet foods) fiber ingredient addition decreases product expansion [31] and increases cutting force [32]. However, when considering the diversity of foods in the grocery stores, there are several examples of insoluble and soluble fibers which have been used successfully in select products [33].

4. Chemical and physical characterization

Before detailing the uses and effects of Miscanthus grass as a fiber source for monogastric animals, it is beneficial to gain an understanding regarding how fiber as a nutrient is characterized. While the term "fiber" is commonly used, it relates to a very diverse group of compounds that are not easy to characterize and quantify. To add to the complexity of this food group, differences in raw material composition (plant variety, age at harvest, environmental conditions, and harvest date) and the process in which the plant material was produced can influence the composition and concentration of the fiber nutrient in the final ingredient [26, 34]. Regardless of the challenges to evaluate fiber sources [35], it is important to characterize the fiber content of an ingredient to properly understand its effects on food processing and the possible health benefits it may have.

Different methods are used across industries to quantify the fiber content of ingredients and foods. Historically, the method initially developed was "crude fiber" (Thaer, 1809 and Hennenburg and Stohmann, 1860 and 1864 in [36]). In this method the sample is digested in a strong acid and then in a base with the residue remaining considered as fiber. In this procedure, all the soluble fibers are washed away; thus, underestimating the total fiber content of the sample. However, this is the method required on the pet food labels by state feed control officials as outlined by Model Bill within the Official Publication for the American Association of Feed Control Officials [37]. Other methods have been developed to measure fiber in forages [38–40] and are common for the beef, dairy, swine, and poultry industries. These procedures boil the forage in neutral or acid detergent solutions and measure the resulting residue. Like the crude fiber method, several of the soluble components of the sample are washed away and not accounted in the measure of fiber. In an attempt to recover the soluble fibers, the total dietary fiber method (TDF) [41] was developed to capture all the fibrous fractions. It was revised a few years later to include the analysis for the insoluble and soluble fractions [42]. This procedure is based on an enzymatic digestion to remove the proteins and starches from the sample. This method is commonly used by the human foods and nutrition industry, as some of its results are correlated with some health benefit. Since some fibers are not recovered by the TDF analysis, other methods have been developed to quantify the fiber content of a given sample; however, they are not standardized and variation in the procedures and results are known to occur [35]. **Table 1** provides a summary of the methods and what fiber component is or not recovered by them. For the sake of this review, fiber composition will be classified by its solubility in water (soluble vs. insoluble) and fermentability (fermentable vs. non-fermentable). We have evaluated the composition of Miscanthus grass as an ingredient for pet food production and its composition is shown on **Table 1**. From the values reported, clearly Miscanthus grass is a source rich in insoluble fibers with some meaningful amount of lignin consistent with most forages.

Method	Fraction Recovered	Unrecovered Fraction	Industry user	Miscanthus grass, %	Wheat bran, %
Crude fiber	Most of the cellulose Some lignin	Soluble fibers, hemicellulose, most of the lignin, and some cellulose	Pet food and Animal feed	45.2	7.5–10.1[1]
Neutral detergent fiber	Cellulose, hemicellulose, lignin	soluble fibers	Animal feed	73.8	23.1–26.5[2]
Acid detergent fiber	Cellulose and lignin	Soluble fibers, hemicellulose	Animal feed	53.7	6.5–8.1[2]
Acid detergent lignin	Lignin	Soluble fibers, cellulose, hemicellulose	Animal feed	13.0	2.4–2.6[2]
Total dietary fiber	Insoluble fibers and most of soluble fibers	Oligosaccharides	Human foods	85.5	33.4–63.0[3]
Insoluble fiber[*]	Insoluble fibers	Soluble fibers	Human foods	78.6	28.4–58.0
Soluble fiber[*]	Most soluble fibers	Insoluble fibers, oligosaccharides	Human foods	6.9	5.0[4]

[*]As part of the total dietary fiber method.
[1]From Food and Agriculture Organization [43].
[2]From Hossain et al. [44].
[3]From Curti et al. [45].
[4]From Babu et al. [46].

Table 1.
Methods commonly used to analyze fiber content of ingredients and values for Miscanthus grass and wheat bran from research referenced in this review.

On the physical side of fiber analysis, the most common analytical method used to characterize ingredients for the production of animal foods is particle size and its distribution. This is usually done with the standard method described by the American Society of Agriculture and Biological Engineers ([47], method S319.4) which consists of stacked sieves in a shaker tapping device. In the procedure a sample is placed on the top sieve and after 10 min on the shaker the content remaining in each subsequent sieve below is weighed and the geometric mean diameter of the particle is calculated from the sieve hole size and residual weight. This is not a characterization of the ingredient as a whole, but rather the specific batch and grinding equipment, as the grind size can be adjusted as needed (**Figure 1F**). For example, in the work of [1] they used a fine (108.57 ± 66.25 μm) and a coarse particle size (294.10 ± 253.22 μm) Miscanthus grass to evaluate the possible effects of particle size in broiler chicken performance and digestibility. This laboratory group has also reported use of a similar fine particle size Miscanthus grass used in a feeding study with cats. In this experiment the particle size of the Miscanthus grass was 103.46 ± 76.39 μm [5] and had positive effects. Pontius et al. [10] reported the exploration of Miscanthus grass as a potential premix carrier. In this work the average particle size was 134 ± 93 μm. They also evaluated flowability and angle of repose (a measure of resistance to flow) of powdered ingredients considered in a manufacturing setting for their ability to move out of bin-bottoms and through transfer pipes [48]. The angle of repose is estimated after a certain amount of the powdered ingredient has been poured onto a level bench top. The lower the angle, the easier the material will flow. The flowability index (FlowDex) is measured by adding a known amount of the powdered ingredient into a cylindrical hopper with a fitted disk of known orifice diameter. The minimum diameter for the material to flow freely is determined after 3 successful tests. From the evaluation of [10] they were unable to determine the flowability index of Miscanthus grass since the ingredient did not flow through the biggest diameter disk (34 mm diameter). Additionally, angle of repose for MG was 47.8° which compared unfavorably to all other tested fibers. These characteristics indicate that Miscanthus grass in a simple ground form may have poor flowability. Though that might be modified with alternative processing steps as has been applied to other fiber carriers and excipients from other sources (*e.g.*, cellulose).

5. Effects on the animal's nutrition and health

As mentioned previously, fiber is not considered an essential nutrient for animals. Although its consumption can be beneficial for reducing energy intake, promoting satiety, supporting gut health, and hairball management [26, 49–55].

Fiber can be of particular interest for the health and wellbeing of cats as they are known to suffer from hairballs. Hairballs, also known as trichobezoars, are hair masses formed in the cat's stomach due to the extensive period of time they groom themselves [54, 56, 57] and some anatomical [57, 58] and physiological adaptations [59]. As a result of these idiosyncrasies, cats can accumulate hair in the stomach and regurgitate it when the mass is too big to pass to the duodenum. In addition, there are reports of intestinal blockages caused by trichobezoars [60]. It is believed that the addition of fiber in the diet can decrease or eliminate this issue. For example, [61] patented (patent number US 7,425,343 B2) the use of high fiber concentrations in the diet for the purpose of improving gastric motility in an effort to pass the trichobezoars to the small intestine and(or) increase the gastrointestinal passage rate. Other fibers have been evaluated as well [5, 54, 62, 63] with variable success. Their inconsistent results may be related to different methodologies used for evaluation of animal responses and the types of fiber used. Clearly, any comparison between studies must be approached with caution and more studies are needed to determine the effects of fiber in hairball management in cats. Miscanthus grass was evaluated as a fiber source to aid in hairball management in cats [5]. In this research trial, 12 American short-hair cats were fed a control diet and a test diet in which Miscanthus grass was added at 10% in exchange of rice flour. The cats were fed the diets for 21 days (16 adaptation days plus 5 days of total fecal collection) with fresh water available throughout the duration of the trial. In addition, cats were brushed prior to the start of each feeding period of a switch-back study design to remove loose hair. It was observed that less hair clumps and total hair weight were excreted per gram of dry feces in cats fed the Miscanthus grass diet. While these results were somewhat expected, because more dry feces was evacuated by cats fed Miscanthus grass, it also provided an indication that fibers (in this case Miscanthus grass) could be used in hairball management in cats as a matter of hair dilution and (or) separation to avoid aggregation. However, it is crucial to state some of the limitations of this trial, such as the use of cats that did not have a history of hairballs and had short hair. Future studies should consider evaluation by cats that have a history of hairballs, have longer hair, and the feeding period should be longer (since regurgitation frequency of a hairball could be monthly) in order to gain a true assessment of hairball elimination.

In similar fashion, weight management, food acceptance, digestibility, fecal consistency and defecation frequency, and colonic fermentation are also affected by the type of fiber. A variety of fiber ingredients are currently used in food production or for supplements intended for both humans and their pets. In general, it is known that obesity can lead to major chronic health issues for humans and pets [53, 64–68]. In theory weight loss by calorie restriction or alternatively an increase in energy expenditure is a simple principle, but in practice it is much more complicated as evidenced by the growing numbers of obese individuals [24] and pets [23]. Dietary fiber ingredients can contribute to caloric restriction and increase the perception of satiety [49, 69]. Unfortunately, dietary fiber addition is also known to decrease acceptance or palatability of a food [27, 70, 71] which contributes to the relatively low success of weight loss/management programs.

Other benefits of fiber in the diet are related to the production of fermentation products in the colon that promote health through the production of post-biotics, especially the short chain fatty acid butyrate. The benefits of butyrate for human health have been extensively reviewed elsewhere [25, 72]; however, there is still the need to verify most of these benefits for pets. The rate of fermentation and the amount of each SCFA is dependent on the fiber source [51, 52, 73, 74]. Thus, if the fiber source is concentrated in soluble and fermentable fibers rather than insoluble and non-fermentable fibers, more SCFA will be produced [75–77]. Miscanthus grass has been evaluated in an in vitro fermentation model using canine feces as an inoculum [3] and its fermentation was comparable to cellulose, an insoluble and non-fermentable fiber source. As a result, Miscanthus grass may not be an effective prebiotic in companion animal diets. Finet et al. analyzed total phenols and indoles, short- and branched-chain fatty acids, and ammonia in fecal samples of cats after they were fed a diet containing 9% Miscanthus grass for 21 days. The authors reported that cats fed Miscanthus grass diet had a higher excretion of indoles compared to cats fed either beet pulp (11% inclusion) or cellulose (7% inclusion). Additionally, acetate and propionate fecal concentrations were also lower compared to cats fed the beet pulp diet; however, no changes in butyrate, branched-chain fatty acids, and ammonia were reported [7]. The addition of Miscanthus grass to feline diet at 9% increased alpha diversity compared to beet pulp supplemented diet when considering Faith's phylogeny and Shannon entropy index [7]. This suggests that while not as substantially fermented compared to other fiber sources, there may be some soluble and fermentable substrate in Miscanthus grass that could benefit the animal if provided at a sufficient dose.

By definition fiber escapes upper gastrointestinal tract digestion and would be available for fermentation in the colon. With more fiber in the diet, dry matter, organic matter, and energy digestibility of foods would decrease [78]. This contributes to dietary energy dilution, especially for insoluble fibers. Dogs [2] and cats [5] fed diets containing 10% Miscanthus grass each had decreased dry matter, organic matter and total dietary fiber digestibility compared to animals fed diets containing a similar level of beet pulp. That [7] did not see an effect of Miscanthus grass (9% inclusion) on dry matter, organic matter, and energy digestibility of dried cat foods compared to those fed diets containing beet pulp is a bit of a mystery. When diets containing 3% Miscanthus grass were fed to broiler chicks, gross energy and apparent metabolizable energy digestibility were lower compared to chickens fed beet pulp diets [1] without changes in dry matter and organic matter digestibility reported. A summary of the digestibility studies published in which Miscanthus grass was a primary fiber source for monogastric animals can be found in **Table 2**.

While this is expected, for some animal industries (*e.g.*, swine and poultry) the addition of fiber is considered to be a nutrient dilution which is undesirable and kept to a minimum. However, there is some indication that addition of fiber ingredients could be beneficial for poultry production and might decrease or replace the use of antibiotics as growth promoters by stimulating the growth of beneficial gut bacteria [80–82]. Further, Miscanthus grass might not qualify as a prebiotic, but its coarse physical characteristics in the feed provided to chicks may stimulate gizzard

Parameter	Chick[1]	Dog[2]	Cat[3]	Cat[4]
Miscanthus grass inclusion, % as is	3.00	10.00	10.00	9.00
Excreta/Feces Dry matter, %	45.25	38.70	34.33	45.93
Defecation frequency, no/day/animal	n/a	2.98	1.25	n/a
Fecal score[5]	n/a	3.64	3.32	3.20
		Digestibility, %		
Dry matter	78.83	78.20	76.20	78.30
Organic matter	79.74	82.10	80.50	81.80
Gross energy	80.52	82.30	81.70	n/a
Crude protein	n/a	87.90	85.80	84.60
Crude fat	n/a	90.70	85.00	91.70
Total dietary fiber	n/a	46.10	20.80	19.10

[1]*From Donadelli et al. [1]; values are averages of tested life stages and the two different tested Miscanthus grass particle sizes.*
[2]*From Donadelli and Aldrich [2].*
[3]*From Donadelli and Aldrich [5].*
[4]*From Finet et al. [7]; fecal scores converted to a similar scale to the other studies.*
[5]*According to Carciofi et al. [79]; 1 = liquid diarrhea, 5 = hard pellets.*
n/a: not available.

Table 2.
Summary of digestibility and stool quality animal studies with Miscanthus grass as a dietary fiber source.

contractions which is known to stimulate digestive secretions. This may improve nutrient digestibility and limit bacterial growth in the proventriculus with hydrochloric acid release [82].

Fiber ingredients can aid fecal consistency and defecation frequency; however, their effects are source and dose dependent [26, 83, 84]. When fed to dogs and cats, the addition of dietary Miscanthus grass did not affect defecation frequency; however, fecal dry matter was higher for animals fed Miscanthus grass [2, 5] compared to pet fed beet pulp. Moreover, feces of dogs and cats fed Miscanthus grass were harder than animals fed beet pulp.

One benefit that Miscanthus grass could have in human health is the control of cholesterol levels. Lignin was shown to have hypocholesterolemic effects in mice [85]. While Miscanthus grass still needs to be evaluated in humans, this could be another use of this fiber source.

6. Effects on food processing and texture

In addition to health, nutrition, and palatability effects, dietary fiber inclusion brings challenges to food processing and texture. As the health food segments expanded in retail stores, so has the number of fiber-added foods and supplements. Common examples of foods that are enriched with fiber include breakfast cereals, bakery goods, pet foods and treats. The two main processes used to manufacture these products are extrusion and baking. In the case of extrusion, fibrous ingredients impact product expansion negatively. Expansion occurs at the end of the die as material is exiting the extruder barrel. At this point there is a pressure difference (inside extruder barrel vs. ambient) which causes the superheated water droplets contained within the starchy matrix to vaporize. This pushes out on the starch matrix which quickly expands to form a foam-like structure. This attribute has been extensively discussed in other publications [31, 86, 87]. During this expansion process there are three key effects fibers have on expansion in these products. First, more dietary fiber means less starch in the formula – starch is the component responsible for the formation of the continuous matrix that expands and creates the product structure. Second, fibrous ingredients may compete with starch for water and limit its [starch] hydration. Third, fibers can disrupt the continuous melt formation (in the case of insoluble fibers) or create weaker melts (when soluble fibers are present). Regardless of the type of fiber, expansion will be impaired as the bubbles formed will prematurely burst [88–90]. As confirmation of this phenomenon, the addition of Miscanthus grass (an insoluble fiber source) decreased radial expansion and increased longitudinal expansion compared to beet pulp (a more soluble fiber source). These differences in how the kibble expanded also impacted sectional expansion ratio index, which was higher for beet pulp diet compared with Miscanthus grass containing food. As the structure is altered due to differences in expansion, Miscanthus grass kibbles required more energy to compress compared to beet pulp kibbles; however, hardness was similar [4]. For the cat foods addition of Miscanthus grass had no effects on tested extrusion parameters or kibble traits [6] compared to cellulose and beet pulp. Conversely, dog foods with Miscanthus grass required less mechanical energy to process compared to beet pulp supplementation [4].

Various fiber sources have been used in human foods at different inclusion levels and for different purposes [91–93]; however, to our knowledge, Miscanthus grass has not been tested for human foods or supplements as of this date.

7. Other Gramineae

Gramineae, or Poaceae, is a family of plants that includes most of the cereal grains (*e.g.*, wheat, rice, corn, sorghum, barley, millet, rye, triticale), bamboos, grasses used for pastures and lawns, and sugarcane for sugar and ethanol production. This is a very diverse family with several uses for humans and animals. Since most of the cereals and the grasses for pastures and lawns are well studied, we will not cover those uses in this chapter. While some bamboo species are used in North America and Europe as an ornamental plant, in Asia, it is a commonly used construction material [94]; however, those uses are beyond the scope of this chapter.

From a nutrition perspective, cereals are an important food source for humans and other monogastric animals. Most commonly, the grains and their various components are used to produce foods for humans and animals. The stalks of the plant are usually left in the fields or burned to produce energy. Another Gramineae largely used by humans is sugarcane. Most of it for the production of sugar and ethanol. Other than these mainstream products limited research is available describing their use in monogastric animals. Specifically, [32] evaluated the use of sugarcane fiber (a co-product of the extraction of the sugarcane juice) as a fiber source for dogs. Compared to wheat bran, sugarcane fiber addition (9% inclusion) decreased the specific mechanical energy necessary to produce the food and increased the cutting force necessary to cut the kibble. When this diet with sugarcane fiber was fed to dogs they preferred the control (no fiber added) diet [27]. As noted previously, this was expected since addition of fiber ingredients generally reduce food palatability.

8. Conclusions and future

As described by different authors, *Miscanthus x giganteus* is a perennial with great potential to be cultivated in cold climates and has good biomass yields. From this crop, Miscanthus grass is produced by simply grinding the dried canes into a powder. This fibrous food ingredient is mostly composed of insoluble fibers with appreciable amounts of lignin, has poor flowability properties, which could bring challenges to a food production facility. Miscanthus grass has been evaluated as a fiber source for dogs, cats, and chicks. There are some benefits to its use through improved chick performance and feed energy utilization. For dogs and cats, it could be used in weight control diets and in hairball management cat foods. Like other fibers, during processing it decreased the expansion of extruded pet foods which may require minor process modifications to effectively achieve product specifications. Based on these findings Miscanthus grass is one of the first forage grasses that have been evaluated as a viable form of supplemental fiber for monogastric animal diets. Whether it will serve a similar purpose in human diets remains to be evaluated, but the potential exists that it might be a viable alternative compared to other fibers currently utilized in the market. What the future holds for Miscanthus grass is uncertain; however, more research is needed to better understand the potential this crop has since its widespread use in animal and human foods could aid in improving health through diet energy dilution, hairball management, and weight management and thereby improve health and wellbeing of animals and people through a well-established and structured supply chain.

Author details

Renan Donadelli and Greg Aldrich*
Department of Grain Science and Industry, Kansas State University, Manhattan, KS, USA

*Address all correspondence to: aldrich4@ksu.edu

IntechOpen

References

[1] Donadelli RA, Stone DA, Aldrich CG, Beyer RS. Effect of fiber source and particle size on chick performance and nutrient utilization. Poultry Science. 2019:98:5820-5830. DOI: http://dx.doi.org/10.3382/ps/pez382

[2] Donadelli RA, Aldrich CG. The effects on nutrient utilization and stool quality of Beagle dogs fed diets with beet pulp, cellulose, and Miscanthus grass. Journal of Animal Science. 2019:97(10):4134-4139. DOI: 10.1093/jas/skz265

[3] Donadelli RA, Titgemeeyer EC, Aldrich CG. Organic matter disappearance and production of short- and branched-chain fatty acids from selected fiber sources used in pet foods by a canine *in vitro* fermentation model. Journal of Animal Science. 2019:97(11):4532-4539. DOI: 10.1093/jas/skz302

[4] Donadelli RA, Dogan H, Aldrich CG. The effects of fiber source on extrusion parameter and kibble structure of dry dog foods. Animal Feed Science and Technology. 2021:274:114884. DOI: https://doi.org/10.1016/j.anifeedsci.2021.114884

[5] Donadelli RA, Aldrich CG. The effects of diets varying in fibre source on nutrient utilization, stool quality and hairball management in cats. Journal of Animal Physiology and Animal Nutrition. 2020:104:715-724. DOI: 10.1111/jpn.13289

[6] Donadelli RA, Dogan H, Aldrich CG. The effects of fiber source on extrusion processing parameters and kibble characteristics of dry cat foods. Translational Animal Science. 2020:4(4):1-8. DOI: 10.1093/tas/txaa185

[7] Finet SE, Southey BR, Rodriguez-Zas SL, He F, de Godoy MRC. Miscanthus grass as a novel functional fiber source in extruded feline diets. Frontiers in Veterinary Science. 2021:8:1-13. DOI: 10.3389/fvets.2021.668288

[8] Anderson E, Arundale R, Maughan M, Oladelnde A, Wycislo A, Volgt T. Growth and agronomy of Miscanthus x giganteus for biomass production. Biofuels. 2011:2(1):71-87. DOI: https://doi.org/10.4155/bfs.10.80

[9] Adams JMM, Winters AL, Hodgson EM, Gallagher JA. What cell wall components are the best indicators for Miscanthus digestibility and conversion to ethanol following variable pretreatments? Biotechnology for Biofuels. 2018:11:67-80. DOI: https://doi.org/10.1186/s13068-018-1066-3

[10] Pontius B, Aldrich CG, Smith S. Evaluation of carriers for use in supplemental nutrient premixes in pet food and animal feeds. In: Proceedings of the Petfood Forum; 23-25 April 2018; Kansas City, MO: PFF, 2018. p. 14.

[11] Clifton-Brown J, Chiang YC, Hodkinson TR. Miscanthus: genetic resource and breeding potential to enhance bioenergy production. In: Vermerris W, editor. Genetic improvement of bioenergy crops. Springer Science & Business Media; 2008. p. 273-294. DOI: https://doi.org/10.1007/978-0-387-70805-8_10

[12] Lewandowski I, Clifton-Brown J, Scurlock JMO, Huisman W. Miscanthus: European experience with a novel energy crop. Biomass Bioenergy. 2000:19:209-227. DOI: https://doi.org/10.1016/S0961-9534(00)00032-5

[13] Visser P, Pignatelli V. Utilization of Miscanthus. In: Jones MB, Walsh M, editors. Miscanthus for energy and fiber. James & James Science Publishers; 2001. p. 109-154. DOI: https://doi.org/10.4324/9781315067162

[14] Cousins AB, Badger MR, Von Caemmerer S. C$_4$ photosynthetic isotope exchange in NAD-ME- and NADP-ME-type grasses. J. Exp. Bot. 2008:59(7):1695-1703. DOI: 10.1093/jxb/ern001

[15] Carroll A, Somerville C. Cellulosic biofuels. Annu. Rev. Plant. Biol. 2009:60:165-182. DOI: 10.1146/annurev.arplant.043008.092125

[16] Beale CV, Long SP. Seasonal dynamics of nutrient accumulation and partitioning in the perennial C$_4$-grasses *Miscanthus × giganteus* and *Spartina cynosuroides*. Biomass Bioenergy. 1997:12(6):419-428. DOI: https://doi.org/10.1016/S0961-9534(97)00016-0

[17] Clifton-Brown j, Lewandowski I. Water use efficiency and biomass partitioning of three different Miscanthus genotypes with limited and unlimited water supply. Annal of Botany. 2000:86:191-200. DOI: 10.1006/anbo.2000.1183

[18] Lewandowski I, Scurlock JMO, Lindvall E, Christou M. The development and current status of perennial rhizomatous grasses as energy crops in the US and Europe. Biomass Bioenergy. 2003:25(4):335-361. DOI: 10.1016/S0961-9534(03)00030-8

[19] Stewart JR, Toma Y, Fernandez FG, Nishiwaki A, Yamada T, Bollero G. The ecology and agronomy of Miscanthus sinensis, a species important to bioenergy crop development in its native range in Japan: a review. Glob. Change Biol. Bioenergy. 2009:1(2):126-153. DOI: 10.1111/j.1757-1707.2009.01010.x

[20] Buhler DD, Netzer DA, Riemenschneider DE, Hartzler RG. Weed management in short rotation poplar and herbaceous perennial crops grown for biofuel production. Biomass and Bioenergy. 1998:14(4):385-394. DOI: https://doi.org/10.1016/S0961-9534(97)10075-7

[21] Bullard MJ, Heath MC, Nixon PMI. Shoot growth, radiation interception and dry matter production and partitioning during the establishment phase of *Miscanthus sinensis 'Giganteus'* grown at two densities in the UK. Annal of Applied Biology. 1995:126(2):365-378. DOI: https://doi.org/10.1111/j.1744-7348.1995.tb05372.x

[22] Prasifka JR, Bradshaw JD, Meagher RL, Nagoshi RN, Steffey KL, Gray ME. Development and feeding of tall armyworm on Miscanthus x giganteus and switchgrass. J. Econ. Entomol. 2009:102(6):2154-2159. DOI: 10.1603/029.102.0619

[23] Association of Pet Obesity Prevention. U.S. Pet obesity survey [Internet]. 2021. Available from: https://petobesityprevention.org/2018

[24] World Health Organization. Obesity and overweight [Internet]. 2021. Available from: https://www.who.int/news-room/fact-sheets/detail/obesity-and-overweight

[25] Hamer HM, Jonkers D, Venema K, Vanhoutvin S, Troost FJ, Brummer RJ. The role of butyrate on colonic function. Alimentary Pharmacology & Therapeutics. 2008:27:104-119. DOI: 10.1111/j.1365-2036.2007.03562.x

[26] Fahey GC, Merchen NR, Corbin JE, Hamilton AK, Serbe KA, Lewis SM, Hirakawa DA. Dietary fiber for dogs: I. Effects of graded levels of dietary beet pulp on nutrient intake, digestibility, metabolizable energy and digesta mean retention time. Journal of Animal Science. 1990:68(12):4221-4228. DOI: 10.2527/1990.68124221x

[27] Koppel K, Monti M, Gibson M, Alavi S, Di Donfrancesco B, Carciofi AC. The effects of fiber inclusion on pet food sensory characteristics and palatability. Animals. 2015:5:110-125. DOI: 10.3390/ani5010110

[28] Sudha ML, Indumathi K, Sumanth MS, Rajarathnam S, Shashirekha, MN. Mango pulp fiber waste: characterization and utilization as a bakery product ingredient. Food Measure. 2015:9:382-388. DOI: 10.1007/s11694-015-9246-3

[29] Kirjoranta S, Knaapila A, Kilpelainen P, Mikkonen KS. Sensory profile of hemicellulose-rich wood extracts in yogurt models. Cellulose. 2020:27:7607-7620. DOI: https://doi.org/10.1007/s10570-020-03300-9

[30] Gomez M, Martinez MM. Fruit and vegetable by-products as novel ingredients to improve the nutritional quality of baked goods. Critical reviews in food science and nutrition. 2018:58(13):2119-2135. DOI: http://dx.doi.org/10.1080/10408398.2017.1305946

[31] Wang S, Kowalski RJ, Kang Y, Kiszonas AM, Zhu MJ, Gajyal GM. Impacts of the particle sizes and levels of inclusions of cherry pomace on the physical and structural properties of direct expanded corn starch. Food Bioprocess and Technology. 2017:10:394-406. DOI: 10.1007/s11947-016-1824-9

[32] Monti M, Gibson M, Loureiro BA, As FC, Putarov TC, Villaverde C, Alavi S, Carciofi AC. Influence of dietary fiber on macrostructure and processing traits of extruded dog food. Animal Feed Science and Technology. 2016:220:93-102. DOI: http://dx.doi.org/10.1016/j.anifeedsci.2016.07.009

[33] Sharma S, Bansal S, Mangal M, Dixit AK, Gupta RK, Mangal AK. Utilization of food processing by-products as dietary, functional, and novel fiber: a review. Critical Review in Food Science and Nutrition. 2016:56:1647-1661. DOI: 10.1080/10408398.2013.794327

[34] Cole JT, Fahey GC, Merchen NR, Patil AR, Murray SM, Hussein HS, Brent JL. Soybean hulls as a dietary fiber source for dogs. Journal of Animal Science. 1999:77(4):917-924. DOI: 10.2527/1999.774917x

[35] Fahey GC, Novotny L, Layton B, Mertens DR. Critical factors in determining fiber content of feeds and foods and their ingredients. The Journal of AOAC International. 2018:101:1-11. DOI: https://doi.org/10.5740/jaoacint.18-0067

[36] van Soest PJ. Symposium on Nutrition and Forage and Pastures: New chemical procedures for evaluating forages. Journal of Animal Science. 1964:23(3):838-845. DOI: https://doi.org/10.2527/jas1964.233838x

[37] Association of American Feed Control Officials (AAFCO). Model Regulations for Pet Food and Specialty Pet Food Under the Model Bill. In: Cook S, editor. AAFCO 2019 Official Publication. Association of American Feed Control Officials, Inc; 2019. p. 139-232.

[38] van Soest PJ. Use of detergent in the analysis of fibrous feeds. II. A rapid method for the determination of fiber and lignin. Journal of the Association of Official Agricultural Chemists. 1963:46:829-835. DOI: https://doi.org/10.1093/jaoac/46.5.829

[39] van Soest PJ, Wine RH. Use of detergents in the analysis of fibrous feeds. IV. Determination of plant cell-wall constituents. Journal of the Association of Official Agricultural Chemists. 1967:50:50-55. DOI: https://doi.org/10.1093/jaoac/50.1.50

[40] van Soesst PJ, Wine RH. Determination of lignin and cellulose in acid-detergent fiber with permanganate. Journal of the Association of Official Agricultural Chemists. 1968:51:780-785. DOI: https://doi.org/10.1093/jaoac/51.4.780

[41] Prosky L, Asp NG, Furda I, DeVries JW, Schweizer TF, Harland BF. Determination of total dietary fiber in food and food products: Collaborative study. Journal of the Association of Official Analytical Chemists. 1985:68(4):677-679.

[42] Prosky L, Asp NG, Schweizer TF, DeVries JW, Furda I. Determination of insoluble, soluble, and total dietary fiber in foods and food products: interlaboratory study. Journal of the Association of Analytical Chemists. 1988:71(5):1017-1023.

[43] Food and Agriculture Organization. Table 45b Proximate composition of commonly used feed ingredients: Energy [Internet]. 1997. Available from: http://www.fao.org/3/w6928e/w6928e1l.htm

[44] Hossain K, Ulven C, Glover K, Ghavami F, Simsek S, Alamri MS, Kumas A, Mergoum M. Interdependence of cultivar and environment on fiber composition in wheat bran. Aust J Crop Sci. 2013: 7(4):525-531.

[45] Curti E, Carini E, Bonacini G, Tribuzio G, Vittadini E. Effect of the addition of bran fractions on bread properties. Journal of Cereal Science. 2013:57:325-332. DOI: http://dx.doi.org/10.1016/j.jcs.2012.12.003

[46] Babu CR, Ketanapalli H, Beebi SK, Kolluru VC. Wheat bran – composition and nutritional quality: a review. Advances in Biotechnology & Microbiology. 2018:9(1):21-27. DOI: 10.19080/AIBM.2018.09.555754

[47] American Society of Agricultural and Biological Engineers (ASABE). Method of determining and expressing fineness of feed materials by sieving (S319.4). 2008.

[48] Taylor MK, Ginsburg J, Hickey AJ, Gheyas F. Composite method to quantify powder flow as a screening method in early tablet or capsule formulation development. AAPS Pharm Sci Tech. 2000:1(3):1-11. DOI: 10.1208/pt010318

[49] Pappas TN, Melendez RL, Debas HT. Gastric distention is a physiologic satiety signal in the dog. Digestive Diseases and Sciences. 1989:24(10):1489-1493. DOI: 10.1007/bf01537098

[50] Fahey GC, Merchen NR, Corbin JE, Hamilton AK, Serbe KA, Hirakawa DA. Dietary fiber for dogs II: Iso-total dietary fiber (TDF) addition of divergent fiber sources to dog diets and their effects on nutrient intake, digestibility, metabolizable energy and digesta mean retention time. Journal of Animal Science. 1990:68:4229-4235. DOI: 10.2527/1990.68124229x

[51] Sunvold GD, Fahey GC, Merchen NR, Reinhart GA. *In vitro* fermentation of selected fibrous substrates by dog and cat fecal inoculum: influence of diet composition on substrate organic matter disappearance and short-chain fatty acid production. Journal of Animal Science. 1995:73:1110-1122. DOI: 10.2527/1995.7341110x

[52] Sunvold GD, Hussein HS, Fahey GC, Merchen NR, Reinhart GA. *In vitro* fermentation of cellulose, beet pulp, citrus pulp, and citrus pectin using fecal inoculum from cats, dogs, horses, humans, and pigs and ruminal fluid from cattle. Journal of Animal Science. 1995:73:3639-3648. DOI: 10.2527/1995.73123639x

[53] Otles S, Ozgoz S. Health effects of dietary fiber. Acta Scentiarum Polonorum, Technol. Aliment. 2014:13(2):191-202.

[54] Loureiro BA, Monti M, Pedreira RS, Vitta A, Pacheco PDG, Putarov TC, Carciofi AC. Beet pulp intake and

hairball fecal excretion in mixed-breed short haired cats. Journal of Animal Physiology and Animal Nutrition. 2017:101(Supplement 1):31-36. DOI: 10.1111/jpn.12745

[55] Carlson JL, Erickson JM, Lloyd BB, Slavin JL. Health effects and source of prebiotic dietary fiber. Current Developments in Nutrition. 2018:2(3):nzy005. DOI: https://doi.org/10.1093/cdn/nzy005

[56] Panaman R. Behavior and ecology of free-ranging farm cats (Felis catus L). Z Tierpsychol. 1981:56:59-73. DOI: https://doi.org/10.1111/j.1439-0310.1981.tb01284.x

[57] Cannon M. Hair Balls in Cats. A normal nuisance or a sign that something is wrong? Journal of Feline Medicine and Surgery. 2013:15:21-29. DOI: 10.1177/1098612X12470342

[58] Weber M, Sams L, Feugier A, Michel S, Biourge V. Influence of the dietary fiber levels on fecal hair excretion after 14 days in short and long-haired domestic cats. Veterinary Medicine and Science. 2015:1:30-37. DOI: 10.1002/vms3.6

[59] De Vos WC. Migrating spike complex in the small intestine of the cat intestine. Am J Physiol. 1993:265: G619-G627. DOI: 10.1152/ajpgi.1993.265.4.G619

[60] Barrs VR, Beatty JA, Tisdall PLC, Hunt GB, Gunew M, Nicoll RG, Malik R. Intestinal obstruction by trichobezoars in five cats. Journal of Feline Medicine and Surgery. 1999:1:199-207. DOI: 10.1053/jfms.1999.0042

[61] Davenport GM, Sunvold GD, Reinhart GA, Hayek MG. Process and composition for controlling fecal hair excretion and trichobezoar formation. Patent number US 7,425,343 B2. 2008.

[62] Dann JR, Adler MA, Duffy KL, Giffard CJ. A potential nutritional prophylactic for the reduction of feline hairball symptoms. The Journal of Nutrition. 2004:134:2124S-2125S. DOI: https://doi.org/10.1093/jn/134.8.2124S

[63] Beynen AC, Middelkoop J, Saris DHJ. Clinical signs of hairballs in cats fed a diet enriched with cellulose. American Journal of Animal and Veterinary Sciences. 2001:6(2):69-72. DOI: https://doi.org/10.3844/ajavsp.2011.69.72

[64] Kealy RD, Lawler DF, Ballam JM, Mantz SL, Nierv DN, Greeley EH, Lust G, Segre M, Smith GK, Stowe HD. Effects of diet restriction on life span and age-related changes in dogs. Journal of the American Veterinary Medical Association. 2002:220(9):1315-1320. DOI: 10.2460/javma.2002.220.1315

[65] German AJ. The growing problem of obesity in dogs and cats. Journal of Nutrition. 2006:136 (7 Suppl):1940S-1946S. DOI: 10.1093/jn/136.7.1940S

[66] Laflamme DP. Understanding and managing obesity in dogs and cats. Veterinary Clinics of North America: Small Animal Practice. 2006:36(6):1283-1295. DOI: 10.1016/j.cvsm.2006.08.005

[67] German AJ, Hervera M, Hunter L, Holden SL, Morris PJ, Biourge V, Trayhurn P. Improvement in insulin resistance and reduction in plasma inflammatory adipokines after weight loss in obese dogs. Domestic Animal Endocrinology. 2009:37:214-226. DOI: 10.1016/j.domaniend.2009.07.001

[68] Thompson SV, Hannon BA, An R, Holscher HS. Effects of isolated soluble fiber supplementation on body weight, glycemia, and insulinemia in adults with overweight and obesity: a systematic review and meta-analysis of randomized controlled trials. The American Journal of Clinical Nutrition. 2017:106:1514-1528. DOI: https://doi.org/10.3945/ajcn.117.163246

[69] Fekete S, Hullar I, Andrasofszky E, Rigo Z, Berkenyi T. Reduction of the energy density of cat foods by increasing their fiber content with a view to nutrients' digestibility. Journal of Animal Physiology and Animal Nutrition. 2001:85: 200-204. DOI: https://doi. org/10.1046/j.1439-0396.2001.00332.x

[70] Sreenath HK, Sudarshanakrishna KR, Prasad NN, Santhanam K. Characteristics of some fiber incorporated cake preparations and their dietary fiber content. Starch. 1996:48(2):72-76. DOI: https://doi. org/10.1002/star.19960480208

[71] Sharif MK, Butt MS, Anjum FM, Nawaz H. Preparation of fiber and mineral enriched defatted rice bran supplemented cookies. Pakistan Journal of Nutrition. 2009:8(5):517-577. DOI: 10.3923/pjn.2009.571.577

[72] Voet D, Voet JG, Pratt CW. Fundamentals of biochemistry – Life at a molecular level. 5th ed. John Wiley & Sons; 2016. 1206 p.

[73] Biagi G, Cipollini I, Zaghini G. *In vitro* fermentation of different sources of soluble fiber by dog fecal inoculum. Veterinary Research Communication. 2008:32(Supplement 1):S335-S337. DOI: 10.1007/s11259-008-9142-y

[74] Guevara MA, Bauer LL, Abbas CA, Berry KE, Holzgaefe DP, Cecava MJ, Fahey GC. Chemical composition, in vitro fermentation characteristics, and in vivo digestibility responses, by dogs to selected corn fibers. Journal of Agricultura and Food Chemistry. 2008:56:1619-1626. DOI: https://doi. org/10.1021/jf073073b

[75] Casterline JL, Oles CJ, Ku Y. 1997. *In vitro* fermentation of various food fiber fractions. J. Agric. Food Chem. 1997:45:2463-2467. DOI: https://doi. org/10.1021/jf960846f

[76] Bosch G, Pellikaan WF, Rutten PGP, van der Poel AFB, Verstegen MWA, Hendriks WH. Comparative in vitro fermentation activity in the canine distal gastrointestinal tract and fermentation kinetics of fiber sources. Journal of Animal Science. 2008:86:2979-2989. DOI: 10.2527/jas.2007-0819

[77] Cutrignelli MI, Bovera F, Tudisco R, D'Urso S, Marono S, Piccolo G, Calabro S. *In vitro* fermentation characteristics of different carbohydrate sources in two dog breeds (German shepherd and Neapolitan mastiff). Journal of Animal Physiology and Animal Nutrition. 2009:93:305-312. DOI: 10.1111/j.1439-0396.2009.00931.x

[78] Kienzle E, Opitz B, Earle KE, Smith PM, Maskell IE. The influence of dietary fiber components on the apparent digestibility of organic matter in prepared dog and cat foods. Journal of Animal Physiology and Animal Nutrition. 1998:79:46-56. DOI: https://doi.org/10.1111/j.1439-0396.1998.tb00628.x

[79] Carciofi AC, Takakura FS, dr-Oliveira LD, Techima E, Jeremias JT, Brunetto MA, Prada F. Effects of six carbohydrate sources on dog diet digestibility and postprandial glucose and insulin response. J. Anim. Physiol. Anim. Nutr. (Berl). 2008:92:326-336. DOI:10.1111/j.1439-0396.2007.00794.x.

[80] Montagne L, Pluske JR, Hampson DJ. A review of interactions between dietary fiber and the intestinal mucosa, and their consequences on digestive health in young non-ruminant animals. Animal Feed Science and Technology. 2003:108:95-117. DOI: 10.1016/S0377-8401(03)00163-9

[81] Amerah AM, Ravindran V, Lentle RG. Influence of insoluble fiber and whole wheat inclusion on the performance, digestive tract development and ileal microbiota

profile of broiler chickens. British Poultry Science. 2009:50(3):366-375. DOI: 10.1080/00071660902865901

[82] Mateos GG, Jimenez-Moreno E, Serrano MP, Lazaro RP. Poultry response to high levels of dietary fiber source varying in physical and chemical characteristics. Applied Poultry Research. 2012:21:156-174. DOI: http://dx.doi.org/ 10.3382/japr.2011-00477

[83] Flickinger EA, Schreijen EMWC, Patil AR, Hussein HS, Grieshop CM, Merchen NR, Fahey GC. Nutrient digestibilities, microbial populations, and protein catabolites as affected by fructan supplementation of dog diets. Journal of Animal Science. 2003:81:2008-2018. DOI: 10.2527/2003.8182008x

[84] McRae MP. Effectiveness of fiber supplementation for constipation, weight loss, and supporting gastrointestinal function: a narrative review of meta-analysis. Journal of Chiropractic Medicine. 2020:19(1):58-64. DOI: https://doi.org/10.1016/j.jcm.2019.10.008

[85] Raza GS, Maukonen J, Makinen M, Nieme P, Niiranen L, Hibberd AA, Poutanen K, Buchert J, Herzig KH. Hypocholesterolemic effect of the lignin-rich insoluble residue of brewer's spent grain in mice fed a high fat diet. Journal of Agricultural and Food Chemistry. 2018:67:1104-1114. DOI: 10.1021/acs.jafc.8b05770

[86] Lue S, Hsieh F, Huff HE. Extrusion cooking of corn meal and sugar beet fiber: effects on expansion properties, starch gelatinization, and dietary fiber content. Ceral Chemistry, 1991:68(3):227-234

[87] Mendonça S, Grossmann MVE, Verha R. Corn bran as a fiber source in expanded snacks. Food Science and Technology. 2000:33(1):2-8. DOI: https://doi.org/10.1006/fstl.1999.0601

[88] Kokini JL, Chang CN, Lai LS. The role of rheological properties in extrudate expansion. In: Kokini JL, Ho CT, Karwe MW, editors. Food extrusion and technology. Marcel Dekker Inc. 1992. p. 631-653. DOI: https://doi.org/10.1080/07373939308916831

[89] Rockey GJ, Plattner B, de Souza EM. Feed extrusion process description. Revista Brasileira de Zootecnia. 2010:39:510-518. DOI: https://doi.org/10.1590/S1516-35982010001300055

[90] Moraru CI, Kokini JL. Nucleation and expansion during extrusion and microwave heating of cereal foods. Comprehensive Reviews in Food Science and Food Safety. 2003:2:147-165. DOI: https://doi.org/10.1111/j.1541-4337.2003.tb00020.x

[91] Massodi FA, Sharma B, Chauhan GS. Use of apple pomace as a source of dietary fiber in cakes. Plant Foods for Human Nutrition. 2002:57:121-128. DOI: https://doi.org/10.1023/A:1015264032164

[92] Cho SS, Samuel P. Fiber Ingredients Food Applications and Health Benefits. CRC Press; 2009. 516 p. DOI: https://doi.org/10.1201/9781420043853

[93] Rosell CM, Santos E. Impact of fibers on physical characteristics of fresh and staled bake off bread. Journal of Food Engineering. 2010:98:273-281. DOI: 10.1016/j.jfoodeng.2010.01.008

[94] Sharma B, Gatto A, Bock M, Ramage M. Engineered bamboo for structural applications. Construction and Building Materials. 2015:81:66-73. DOI: http://dx.doi.org/10.1016/j.conbuildmat.2015.01.077.

Chapter 6

Top Dressing of Fertilizers: A Way Forward for Boosting Productivity and Economic Viability of Grasslands

Tessema Tesfaye Atumo, Milkias Fanta Heliso,
Derebe Kassa Hibebo, Bereket Zeleke Tunkala
and Yoseph Mekasha

Abstract

Grasslands in the Ethiopian highlands have been degrading with grazing loads. Fertilizers like nitrogen, phosphorus and sulfur improves the soil fertility and species composition of the grazing lands. This study justifies, evaluation of top dressing nitrogen and phosphorus fertilizers on biomass yield of grass lands for market-oriented livestock production studied at Chosha kebele, Southern Ethiopia in 2017. Three fertilizer levels ((T1), 150 kg ha^{-1} urea (T2) and combination of 110 kg ha^{-1} urea and 100 kg ha^{-1} NPS (T3)) were laid out in randomized complete block design with 6 replications in summer and winter cropping seasons. Dry matter yield was significantly (P<0.001) different among treatments and higher results were obtained for combination of urea and NPS, followed by urea and the control one. Higher grasses species composition between application of combination of urea and NPS than urea alone. Net revenue is higher in nitrogen alone application than nitrogen and phosphorus. Therefore, better marginal rate of return (MRR=828%) recorded in Urea application for grazing land improvement in Gamo highland areas. It is recommendable to apply 150 kg/ha urea fertilizer to bring optimum yield of grazing land in Southern Ethiopian Highlands.

Keywords: Nitrogen, Phosphorus, grazing land, dry matter, species composition

1. Introduction

Sub-Saharan livestock production is increasingly constrained by feed shortage, both in quantity and quality [1]. Livestock production can be improved through good management of natural grasslands and introduction of improved fodder species [2] with the supply of fertilizer and water to maintain high productivity that the high cost and low availability of good quality animal feed is a critical constraint to increasing productivity of livestock in dairy farms and feedlots, improved family and specialized poultry, and smallholder mixed crop-livestock and extensive livestock production systems [1].

Nutrient dynamics in tropical soils sustaining forage grasses are still poorly understood [3]. Lack of nutrients, inadequate management of pastures, and inappropriate cultural practices are responsible for pasture degradation. Applying fertilizers in large quantities increase the productivity of grasslands [4]. Low nitrogen availability has been identified as a major cause of degradation of tropical pastures [5] and the constant removal of forage without proper supply of nutrients extracted by plants emphasizes the problems of grazing land degradation [6]. The application of nitrogen and phosphorus has proved to be effective in maximizing the production of dry matter [7] and nutritional status [8] of grasses.

Grazing lands in Ethiopia play great role in livestock production. However, grazing land degradation in Ethiopia is a serious problem [9]. Since a few decades ago, the country is not only known for the severity of grazing land degradation and related problems, but also for concerted efforts to confront the problems using land rehabilitation measures such as enclosures [10]. Enclosures have been widely established particularly in the midland and highland agro-ecologies. They are among the green spots with considerable species diversity and higher biomass production compared the unclosed areas [11] .

Nitrogen and phosphorus fertilizers have been used for long period of time in agricultural system [12]. Nitrogen fertilizer application improves above ground biomass of any plant crops [13]. Phosphorus improves the growth of legumes and plant species composition generally and a Poaceae pasture in specific [14]. Nitrogen and phosuphorus fertilizers combined application could improve the aboveground plant biomass [15] and have positive effects on composition diversity of plant species [16]. Primary mineral fertilizers such as nitrogen, phosphorus, and sulfur etc. are favoring the growth of plants through improving soil fertility [17]. Though enclosures produced better biomass than the freely grazed areas, production is still limited. This probably is because of limited plants growth related to nutrient deficiency. Addition of nitrogen and sulfur fertilizers increased shoot dry matter production in the second and third growth of forage plants [3]. Nitrogen availability maximizes plant growth and productivity [6]. Nitrogen deficiency in the grazing areas of Ethiopian highlands due to land degradation was, which probably could be the leading constraint for limited plant growth and reduced biomass yield, affecting crop production [18]. Hence, application of nitrogen and phosphorus with sulfur seems imperative to enhance plant growth and increase herbage biomass production. The first and foremost beneficiaries of the research findings are small holder farmers, policy makers, researchers and NGOs. The hypothesis of the study was fertilizing grazing lands improve the herbage biomass and economic feasibility and applicability of grazing lands under small holder farmers condition. Therefore, this study was planned to evaluate top dressing of grazing lands in terms of biological gain and assess economic gain and the applicability of pasture fertilization under smallholder private or communal grazing lands.

2. Materials and methods

2.1 Study area

The study was conducted at the highland of Chosha kebele, Bonke district, Gamo Gofa zone, Southern Ethiopia (**Figure 1**). The altitude of the area is 2350 meter above sea level with annual average rainfall of 2017.06 mm and mean daily temperature ranging between 10.0–23.3°C (**Figure 2**). The rainfall is bi-modal with the winter rain (short rains) occurring in March to May and the summer (main season) rains lasting from June to October. Major crops such as potato, wheat, barley,

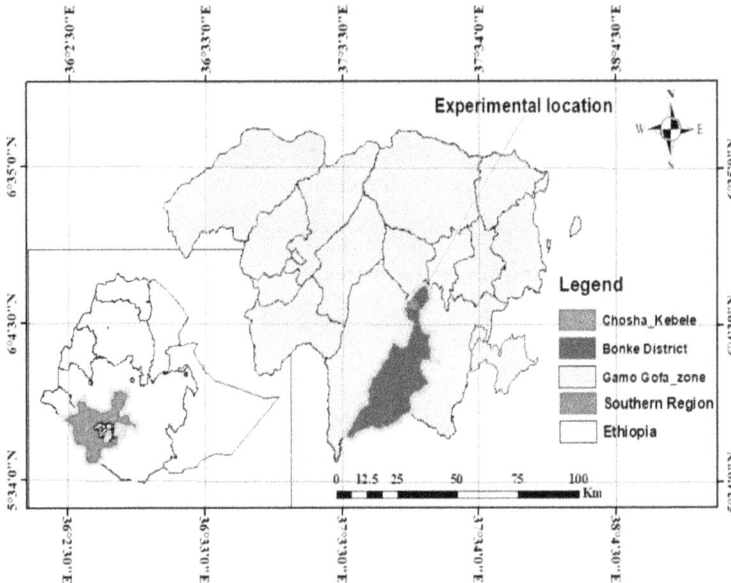

Figure 1.
Location map of experimental site.

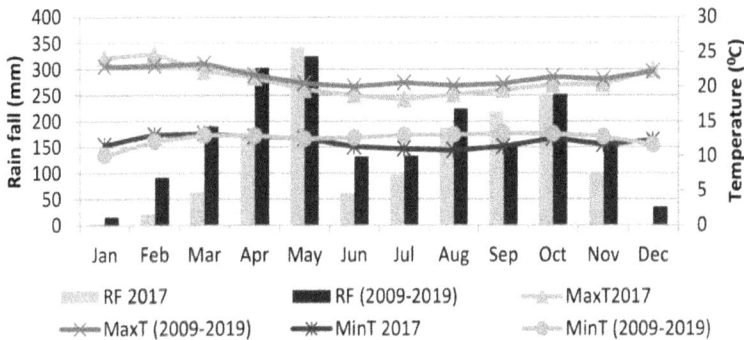

Figure 2.
Rainfall, maximum and minimum temperature of experimental season and ten years average in the location.

bean, onion, paper, cabbage, fruits are grown widely in the study area according to site observation and district report. Natural pasture is the major feed source in the area and farmers using cut and carry system of livestock feeding mainly because of shortage of farming and grazing land. The soil of study area is characterized as strongly acidic with pH < 5.0, low organic carbon contents which ranged from 0.25% to 1.05%, moderate calcium carbonate with 0.88%, high organic matter with 13.56%, low catain exchange capacity with 16.69 cmolc/kg and sandy-loam [19].

2.2 Treatments and experimental design

The fertilizer treatments for the study were T1 = control, T2 = urea and T3 = combined urea and NPS in both summer and winter major cropping seasons of Ethiopia. The amount of urea and NPS that were used in the experiment was 150 kg; and 110 and 100 kg per ha for T2 and T3, respectively. Factorial combination of two seasons and 3 fertilizer treatments laid out in RCBD with four replications. The plot size consisted of an area of 400 m² (20m x 20 m) and the space between plots was 3 m. Nitrogen fertilizer applied in the form of urea as a split dressing i.e., one-third at about 7 days of the first rain and two-thirds after about a month of the first rain and P fertilizer in the form of NPS applied at about 7 days of the first rain together with the nitrogen applied at 7 days after the first rain. The trial is replicated in winter and summer seasons with fertilizer application in March for winter and in June for summer. The fertilizers applied manually in the field determined for experiment.

2.3 Data collection and sampling procedures

2.3.1 Forage yield

Herbage biomass was measured as the herbaceous vegetation harvested at ground level using manual sickle from five 0.5 m quadrates (four at the corner and one at the center of the 10 m x10m plots) using sickle in each of the 100m² plots. Fresh biomass weighed immediately using weighing scale of 0.1 g. Then, a sub-sample of 15–20% of the total weight was separated and put into a paper bag for dry matter determination and oven dried at 105°C for 24 hours.

2.3.2 Species composition

Species composition was determined by using quadrate count method and identified in the field with farmers for local name and taxonomic classification. Species that were difficult to identify in the field recorded and collected to herbarium for identification.

2.3.3 Economic considerations

Partial budget analysis was performed to evaluate the economic advantage of fertilization by using the procedure of Upton (1979). The partial budget analysis involves calculation of the variable costs and benefits. The benefits are calculated based on market value of green or cured grass for all expenses recorded at the beginning of the study.

The amount of herbage obtained used to calculate the income earned (TR). The calculation of the variable costs and the expenditures incurred on various activities were taken into consideration.

The partial budget method measured profit or losses, which is the net benefits or differences between gains and losses for the proposed change and includes calculating net return (NR), i.e., the amount of money left when total variable costs (TVC) are subtracted from the total returns (TR):

$$NR = TR - TVC \qquad (1)$$

Total variable costs included the costs of all inputs that change due to the change in production technology. The change in net return (ΔNR) calculated by the difference between the change in total return (Δ TR) and the change in total variable cost (Δ TVC), and this is used as a reference standard for decision on the adoption of a new technology.

$$\Delta NR = \Delta TR - \Delta TVC \qquad (2)$$

The marginal rate of return (MRR) measured the increase in net income (Δ NR) associated with each additional unit of expenditure (Δ TVC). This is expressed by percentage

$$MRR\% = \frac{\Delta NR}{\Delta TVC} * 100 \qquad (3)$$

2.3.4 Statistical analyses

The experimental data was subjected to analysis of variance using the General Linear Model Procedure of Genstat statsitcal software [20]. Tukey HSD test applied for mean comparisons and statistically significant differences were accepted at $P < 0.05$.

3. Results and discussions

3.1 Herbage biomass

For the present experiment dry matter yield significantly ($P < 0.001$) varied among treatments and higher results were obtained from combination of urea and NPS followed by urea than the control one (**Figure 3**). This may be due to the application of nitrogen, phosphorus and sulfur in the form of urea and blended NPS fastened the growth of grasses, legumes and other species. Nitrogen Fertilizer application increased dry matter yield. Dry matter yield in summer was by far greater than in winter (**Figure 3**) that may be due to moisture stress in winter season which could demonstrate that the growth of pastures improved in rainy season than dry. Dry matter accumulation is physiological index related to photosynthesis of leaves in which legumes respond less to N than grasses; grass dominant pastures well responded to N [21]. The increase in the proportion of grass reflects the role of nitrogen fertilizer in influencing the grass-legume botanical composition in favor of

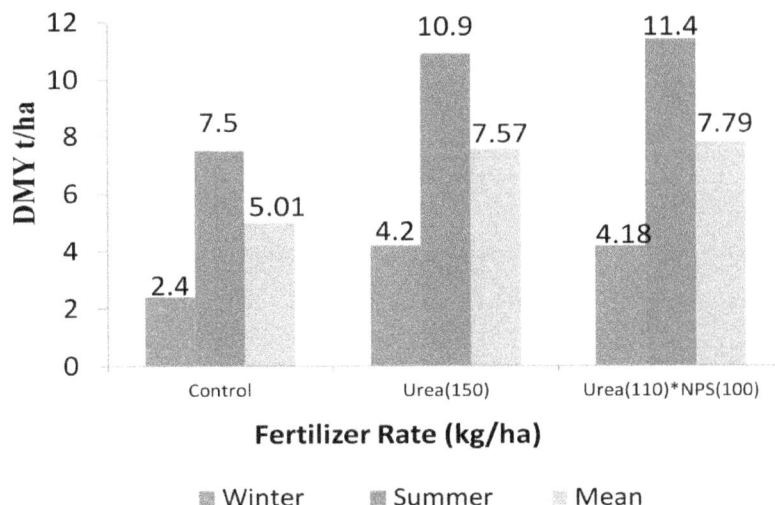

Figure 3.
Dry matter yield (t/ha) as affected by fertilizer application.

grass growth. NPS fertilizer application improved the dry matter yield production of Napier grass in Ethiopia [7] report is in line with the present study. Nitrogen and Phosphorus fertilizers are vital to plant growth and found in every living plant cell and total dry matter yield increment due to nitrogen, phosphorus and potassium fertilizers application reported previously for desho grass production [22]. Nitrogen and phosphorus fertilizers application also improved the growth and crud yield of cauliflower [23]. And also another similar report stated that proper nitrogen and sulfur fertilizer application promotes grass production by improving uptake of nutrients and the dynamics of the organic and mineral fractions in tropical soil [3].

3.2 Species composition

Species composition of the study was presented in **Table 1**. Higher grasses species composition was obtained for combined application of urea and NPS than urea alone. Species composition was being higher for fertilizer application in both levels than control. The difference in species composition of the natural pastureland recorded in this trial is a desirable attribute in terms of pasture quality, quantity and persistence. Hence, the presence of various fodder species in this study would indicate the degree of persistence of some species against recurrent drought, frost and high pasture pressure consistent with the harshness of the prevailing climatic and biotic factors. Application of nitrogen and phosphorus activated growth and development of grasses, legumes and other pastures. Thus, the composition pastures in this study significantly higher in fertilized plots than in control (**Figure 4**). A total of 15 grass species recorded in 9 families with Polygonaceae and Asteraceae taking the highest record and others like Apiaceae the least. Some species like *Bidens macroptera* (Sch.Bip.ex Chiov.) was being very importantly chosen by women of the area for lactating cows. This result invites further immediate investigation of the particular grass species correlation with the milk production and quality. Natural

Local name	Scientific/botanical name	Family	Types
	Agrocharis melanantha Hochst.	Apiaceae	Legume
	Dicrocephala integrifolia (L.f) Kuntze	Asteraceae	
Gocha	Bidens macroptera (Sch.Bip.ex Chiov.)	Asteraceae	Legume(for milk production)
	Gnaphalium rubriflorum Hilliard	Asteraceae	Legume
	Commelina sp.	Commelinaceae	
Gichola	Cyprus triceps Endj,	Cyperaceae	
Donaka	plectranthus punctatus (L.f) L'Her.	Lamiaceae	Legume
Basmamo	Salvia nilotica Jacq.	Lamiaceae	Legume
Dhadhaho	Plantago palmata Hook.f.	Plantaginaceae	Legume
Suda	*D. abyssinica* (Hochst, ex A. Rich.) Stapf	Poaceae	Grass
Hopho	Rumex abyssinicus Jacq.	Polygonaceae	
Shodo	Rumex nepalensis Spreng.	Polygonaceae	
	Persicaria setosula (A. Rich.) K.I. Wilson,	Polygonaceae	Legume
	Alchemilla sp.	Rosaceae	
Tri-folium			Legume

Table 1.
Species' composition identified in the study area, 2017.

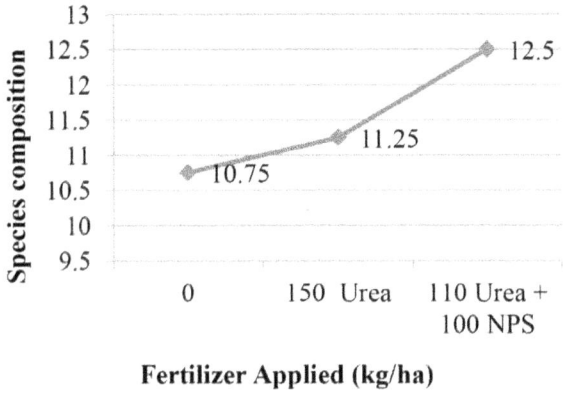

Figure 4.
Response of grass lands on species composition to fertilizer application.

grasslands rich in species composition [24] and fertilizing improves the growth and development of dominated species that higher species composition across all the treatments including the urea applied plots. It was also reported that use of fertilizers increasing plant biomass production and biodiversity in semi-arid grasslands [14]. The average legume proportion was higher in the unfertilized plots than in the fertilized plots and this may indicate nitrogen fertilizer had an indirect suppressing effect on the proportion of legumes by inducing luxuriant growth and hence dominance of the grasses.

3.3 Cost benefit analysis

The partial budget analysis presented in **Table 2** conducted as cost of variable entities was calculated based on cost of fertilizers (Urea and NPS). However, the cost of management like fencing, harvesting, transporting and different activities

Fertilizer kg ha-1			
Descriptions	0	Urea 150	110Urea + 100NPS
Fixed Costs			
Fencing	1.85	1.85	1.85
Harvesting	2.08	2.08	2.08
Total Fixed costs(TFC)	3.93	3.93	3.93
Variable Costs			
NPS(0.29 USD/kg)	0		29.20
Urea(0.24 USD/kg)	0	35.84	26.28
Total Variable Costs(TVC)	0	35.84	55.48
Dry Matter Yield(t/ha)	5.01	7.57	7.79
Total Revenue(TR = 29.97 USD/t)	150.15	226.88	233.47
Net Revenue(NR = TR-TVC)	150.15	191.04	177.9
MRR%(ΔNR/ΔTVC*100)		533.09%	320.82%
MRR- Marginal Rate of Return.			

Table 2.
Partial budget analysis.

disposed for pasture was not included in calculation of variable costs. The price of one ton in the local area was 29.5 USD. Dry matter yield was increasing from control to different fertilizer application and each ton increment in yield influencing the income driving from the production. Net revenue is higher in nitrogen alone application than nitrogen and phosphorus. Therefore, this report verifies better marginal rate of return (MRR = 828%) recorded in Urea application for pastureland improvement in Gamo Gofa highland areas.

4. Conclusion

Dry matter production was higher for combination of urea and NPS followed by urea than control one. Higher grasses species composition between application of combination of urea and NPS than urea alone. Higher net revenue was obtained in nitrogen alone than combined application of nitrogen and phosphorus fertilizers. Therefore, better marginal rate of return (MRR = 828%) recorded in Urea application for pastureland improvement in Gamo Gofa highland areas. Applying nitrogen to pasture land improves dry matter yield in 34% comparing to not applying. A farmer can have net revenue of more than 191.04 USD per hectare on average per season and it is economical to apply nitrogen for pasture land improvement. It is recommended to apply 150 kg/ha urea fertilizer to fetch optimum economical yield of pasture land in southern Ethiopia highlands.

Acknowledgements

The author is grateful for the financial support provided by the Livestock and Irrigation Value Chain for Ethiopian Smallholders (LIVES) project and Southern Agricultural Research Institute (SARI) Livestock Research Directorate to undertake the experiment. His special gratitude also goes to forage crops research colleagues at Arba Minch Agricultural Research Center (AMARC) for their technical and material support throughout the entire work.

Author details

Tessema Tesfaye Atumo[1*], Milkias Fanta Heliso[1], Derebe Kassa Hibebo[1],
Bereket Zeleke Tunkala[2] and Yoseph Mekasha[3,4]

1 Arbaminch Research Center, Southern Agricultural Research Institute (SARI),
Arbaminch, Ethiopia

2 Faculty of Veterinary and Agricultural Sciences, University of Melbourne,
Victoria, Australia

3 Agricultural Transformation Agency (ATA), Addis Ababa, Ethiopia

4 Livestock and Irrigation Value Chain for Ethiopian Smallholders (LIVES),
Addis Ababa, Ethiopia

*Address all correspondence to: tessema4@gmail.com

IntechOpen

References

[1] Shapiro, B. I., Gebru, G., Desta, S., Negassa, A., Negussie, K., Aboset, G., & Mechal, H. (2015). Ethiopia livestock master plan: Roadmaps for growth and transformation. In *International Livestock Research Institute (ILRI)*. https://doi.org/10.11648/j.aff.20140303.11

[2] Mengistu, A., Kebede, G., Assefa, G., & Feyissa, F. (2016). Improved forage crops production strategies in Ethiopia: A review. Academic Research Journal of Agricultural Science and Research, 4(6), 285-296. https://doi.org/10.14662/ARJASR2016.036

[3] Bona, F. D. De, & Monteiro, F. A. (2010). Nitrogen and Sulfur Fertilization and Dynamics in a Brazilian Entisol under Pasture; Nitrogen and Sulfur Fertilization and Dynamics in a Brazilian Entisol under Pasture. Soil Fertility & Plant Nutrition, 74(4). https://doi.org/10.2136/sssaj2009.0228

[4] Ghosh, P. K., Mahanta, S. K., & Ram, S. N. (2017). Nitrogen Dynamics in Grasslands. In *The Indian Nitrogen Assessment: Sources of Reactive Nitrogen, Environmental and Climate Effects, Management Options, and Policies* (pp. 187-205). Elsevier. https://doi.org/10.1016/B978-0-12-811836-8.00013-6

[5] Nyameasem, J. K., Reinsch, T., Taube, F., Yaw Fosu Domozoro, C., Marfo-Ahenkora, E., Emadodin, I., & Malisch, C. S. (2020). Nitrogen availability determines the long-term impact of land use change on soil carbon stocks in grasslands of southern Ghana. Soil, 6(2), 523-539. https://doi.org/10.5194/soil-6-523-2020

[6] Kiba, T., & Krapp, A. (2016). Plant nitrogen acquisition under low availability: Regulation of uptake and root architecture. Plant and Cell Physiology, 57(4), 707-714. https://doi.org/10.1093/pcp/pcw052

[7] Atumo, T. T., Kalsa, G. K., & Dula, M. G. (2021). Effect of Fertilizer Application and Variety on Yield of Napier Grass (*Pennisetum purpureum*) at Melokoza and Basketo Special Districts, Southern Ethiopia. J. Agric. Environ. Sci., 6(1), 32-39. https://journals.bdu.edu.et/index.php/jaes/article/view/458

[8] Maleko, D., Mwilawa, A., Msalya, G., Pasape, L., & Mtei, K. (2019). Forage growth, yield and nutritional characteristics of four varieties of napier grass (Pennisetum purpureum Schumach) in the west Usambara highlands, Tanzania. *Scientific African*, 6. https://doi.org/10.1016/j.sciaf.2019.e00214

[9] Bewket, W. (2007). Soil and water conservation intervention with conventional technologies in northwestern highlands of Ethiopia: Acceptance and adoption by farmers. Land Use Policy, 24(2), 404-416. https://doi.org/10.1016/j.landusepol.2006.05.004

[10] Yami, M., Gebrehiwot, K., Stein, M., & Mekuria, W. (2006). *Impact of Area Enclosures on Density, Diversity, and Population Structure of Woody Species: the Case of May Ba' Ati-Douga Tembien, Tigray,. February 2006.*

[11] Yayneshet, T., Eik, L. O., & Moe, S. R. (2009). The effects of exclosures in restoring degraded semi-arid vegetation in communal grazing lands in northern Ethiopia. Journal of Arid Environments, 73(4-5), 542-549. https://doi.org/10.1016/j.jaridenv.2008.12.002

[12] Lu, C., & Tian, H. (2017). Global nitrogen and phosphorus fertilizer use for agriculture production in the past half century: Shifted hot spots and nutrient imbalance. Earth System

Science Data, *9*(1), 181-192. https://doi.org/10.5194/essd-9-181-2017

[13] Liu, W., Jiang, L., Hu, S., Li, L., Liu, L., & Wan, S. (2014). Decoupling of soil microbes and plants with increasing anthropogenic nitrogen inputs in a temperate steppe. Soil Biology and Biochemistry, *72*, 116-122. https://doi.org/10.1016/j.soilbio.2014.01.022

[14] Tong, Z., Quan, G., Wan, L., He, F., & Li, X. (2019). The effect of fertilizers on biomass and biodiversity on a semi-arid Grassland of Northern China. Sustainability (Switzerland), *11*(10). https://doi.org/10.3390/su11102854

[15] Veresoglou, S. D., Voulgari, O. K., Sen, R., Mamolos, A. P., & Veresoglou, D. S. (2011). Effects of nitrogen and phosphorus fertilization on soil pH-Plant productivity relationships in upland Grasslands of Northern Greece. Pedosphere, *21*(6), 750-752. https://doi.org/10.1016/S1002-0160(11)60178-1

[16] Humbert, J. Y., Dwyer, J. M., Andrey, A., & Arlettaz, R. (2016). Impacts of nitrogen addition on plant biodiversity in mountain grasslands depend on dose, application duration and climate: A systematic review. In *Global Change Biology* (Vol. 22, Issue 1, pp. 110-120). Blackwell Publishing Ltd. https://doi.org/10.1111/gcb.12986

[17] Kramberger, B., Podvršnik, M., Gselman, A., Šuštar, V., Kristl, J., Muršec, M., Lešnik, M., & Škorjanc, D. (2015). The effects of cutting frequencies at equal fertiliser rates on bio-diverse permanent grassland: Soil organic C and apparent N budget. Agriculture, Ecosystems and Environment, *212*, 13-20. https://doi.org/10.1016/j.agee.2015.06.001

[18] Sinore, T., Kissi, E., & Aticho, A. (2018). The effects of biological soil conservation practices and community perception toward these practices in the Lemo District of Southern Ethiopia.

International Soil and Water Conservation Research, *6*(2), 123-130. https://doi.org/10.1016/j.iswcr.2018.01.004

[19] ATA. (2016). *Soil Fertility Status and Fertilizer Recommendation Atlas of the Southern Nations, Nationalities and Peoples' Regional State,* Ethiopia (Vol. 1).

[20] Payne, R., Murray, D., Harding, S., Baird, D., & Soutar, D. (2015). *Introduction to Genstat® for WindowsTM* (18th ed.). VSN International, 2 Amberside, Wood Lane, Hemel Hempstead, Hertfordshire HP2 4TP, UK.

[21] Valentin, K. M., Aliou, S., & Augustin, S. B. (2014). Response to fertilizer of native grasses Response to fertilizer of native grasses (Pennisetum polystachion and Setaria sphacelata) and legume (Tephrosia pedicellata) of savannah in Sudanian Benin. *Agriculture, Forestry and Fisheries*, *3*(3). https://doi.org/10.11648/j.aff.20140303.11

[22] Heliso, M. F., Hibebo, D. K., Atumo, T. T., Tunkala, Z., & Dula, M. G. (2019). Evaluation of Desho Grass (Pennisetum Pedicellatum) Productivity under Different Fertilizer Combinations and Spacing at Gamo Gofa Zone, Ethiopia. *J. Agric. Environ. Sci*, *4*(1), 50-59. https://orcid.org/0000-0001-6347-7058

[23] Sahito, A., Laghari, M. H., Agro, A. H., Hajano, A. A., Kubar, A., Khuhro, W. A., Laghari, F. R., Gola, A. Q., & Wahocho, N. A. (2018). *Effect of various leveles of nitrogen phosphorus on plant growth and curd yield of Cauliflower (Brassica oleraceae L .)*. 8(03).

[24] Blumetto, O., Scarlato, S., Castagna, A., Tiscornia, G., Ruggia, A., & Cardozo, G. (2015). Improving livestock production assuring natural grassland ecosystem conservation: three key management practices at farm level. *XXIII International Grassland Congress*, 3.

Chapter 7

Implement and Analysis on Current Ecosystem Classification in Western Utah of the United States & Yukon Territory of Canada

YanQing Zhang and Neil E. West

Abstract

The study cases in western Utah of the United States and Yukon Territory of Canada have more natural land and conservative ecosystems in North America. The ecosystem classification of land (ECL) in these two ecoregions had been analyzed and validated through implementation. A full ECL case study was accomplished and examined with eight upper levels of ECOMAP plus ecological site and vegetation stand in Western Utah, the US. Theoretically, applying Köppen climate system classification, Bailey's Domain and Division were applied to the United States, North America, and world continents. However, Canada's continental upper level ecoregion framework defined the ecological Mozaic on a sub-continental scale, representing an area of the hierarchical ecological units characterized by interactive and adjusting abiotic and biotic factors. Using Bailey's Domain as the top level of Canada's territorial ecoregion was recommended. Eight levels of ELCs were established for Yukon Territory, Canada. Thus, the second study case recommends integrating the ecosystem approaches with Bailey's upper level ECL, broad ecosystem classification, and objectively defined ecological site in different countries, or ecoregions. Our study cases had exemplified the implementations with a full ELCs in Bailey's 300 Dry Domain and 100 Polar Domain.

Keywords: Ecosystem Classification of Land (ECL), Ecoregion, Hierarchy, Board Ecosystem, Objective Approach, Ecological Site, Dry Domain, Polar Domain

1. Introduction

The ecosystem classification of land is about the theory and design of the ECL framework and implements and practices in different nations, continents, and global scales. Bailey had made his primary studies and contributions on ecological classification framework and application, representing his scientific collections of mapping on ecosystem classification of land for the United States, North America, and global continents in [1, 2]. The ecological sites were studied and monitored with environmental conditions, biological characters, and ecosystem services [3–6]. Ecologists and geographers had proposed and classified the land into simplified ecosystems where the different plants, animals, and bacteria populations lived together. By processing into different scales, geographers and ecologists designed ECL framework, theory, and applications to depict the ecosystem as systemically organized, nested, and multiple layers in [7–9]. They are so complex and adapted a cycle crossing a threshold from one stable state to another depending on the seasonality, time, landscapes, and disturbances in Refs. [10, 11], which results in the academic argument where to draw a line based on prior selected criteria, how to identify ecological sites and classify the ecoregions in Refs. [1, 3, 8, 12–14]. Afterward, do we achieve our research goal?

From a philosophical perspective, ecological regionalization could be concerned as an objective that has a form with a perceptive logic; at other times, it is an inductive and subjective art that reflects a management consideration, which is dependent on the application of the ecoregion. However, with the ecological regionalization, the contributions of existing ecoregion schemes are inconsistent. In other words, it is getting study complete with errors remaining in [11, 15].

A large amount of vector or raster formats data made the quantitative and spatial analysis more useful and practical in the last two decades. The tree technique was used to explore the analysis of complex ecological data with nonlinear relationships and high-order interaction in 2000 [16]. Many studies and attempts to analyze the complex system of nature as dynamically organized and structured within and across the scales of space and seasonality had assisted ecological researchers to solve population richness and dynamics in [17], vegetation distributions in [18, 19], and ecosystem classification framework in Refs. [1, 2, 9, 14, 20–24]. Understanding how environmental variables influenced the vegetation pattern and distribution and successional order, many research works demonstrated a hierarchical paradigm in Refs. [1, 11, 15, 25].

From 1976 to 1998, Bailey started to identify the ecoregion boundaries and generated the ecoregions of the United States, North America, and the world's continents. He published his research works and had made significant progress in the 1990s. In 1993, Bailey classified the ecoregion into the top three level classes: Domain, Division, and Province. Then, applying the Köppen climate system of classification, he depicted the Domains with the synthetic description of the land surface form, climate, vegetation, soils, and fauna, seeing in [1–3]. Since Federal Geographic Data Committee (FGDC) in the United States accepted the National Hierarchy of Ecological Units (NHEU), ECOMAP in [26] was created with eight levels hierarchical approach to study the ecosystem classification of Land (ECL).

Bailey and Jensen published their work on the design and ecological mapping units with nine levels [27]. The Subregions below the Domain, Division, and Province were divided into Sections, Landtype Association, Landtype, Landtype Phase, and Ecological Site. Thus, NHEU and Bailey had driven a classified Ecosystem Classification of Land into the nested hierarchies at various scales, depending on management needs.

In the global context of ecosystem classification of land, we need to understand the landscape-scale processes more generally. The issue focuses on generalizing ecoregions, the landscape-scale variation, and the combination of abiotic and biotic factors. It had been extended to identify the circumstances in which generalizations can be made, where there are limits, and find a solution in Refs. [9, 10, 14, 24, 28, 29]. It was valuable to examine the hierarchies of ecosystem classification of Land {ECL} globally when we had working experiences and research cooperation that can be related in different countries or continents in Refs. [12, 14, 19, 30]. More recently, the ecosystem services and values have been concerned with the wise use of biodiversity and natural resources [6].

In this chapter, we tried to compare the current two national ecosystem classification frameworks and assess any Domain related issue when it existed. We tried to find suitable abiotic and biotic factors, topographic features, climatic, and ecosystem services to generate deliverable lower-level ecosystem classification when these related research works were reported and published. However, this inconsistency in terminology is often confusing because similar terms may have different meanings or apply to different scales, and different terms may have the same meaning in [15]. Therefore, we will stick to our current references and literature for reviewing and discussing.

Two sets of ecoregions data of Western Utah of the United States, Yukon Territory of Canada were analyzed and validated. The Biogeoclimatic Ecosystem Classification (BEC) approach was referred to as an additional assessment in the discussion. Our focus was tried to explore lower level ecosystem classification in the different ecoregions of North America in Refs. [1, 2, 31–39].

2. Methodology and analysis

2.1 The review of upper level ecoregions of the United States

The ecosystem can be a complex system more than we thought, which is changed and varied along with longitude, latitude, and elevation on the earth's surface, and constantly adapted to the slope, aspect, environmental variables in macroscales [1, 2, 7, 9, 15, 17, 24]. Bailey had contributed to the ecological classification framework and application, which represented his scientific collections of mapping on ecosystem classification of the United States (**Figure 1A**).

Theoretically, Bailey's Ecosystem Classification of Land had explained the ecoregions and their nested structures in the upper levels of Domain, Division, and province. However, these advantages had not been fully applied and examined as ECL's bases for Terrestrial Ecozones and Ecoregions of Canada in [31, 36–39], even though technically Bailey's ECL polygons in the upper three levels can be easily retrieved in GIS spatial model in [14] when the ECL project was conducted.

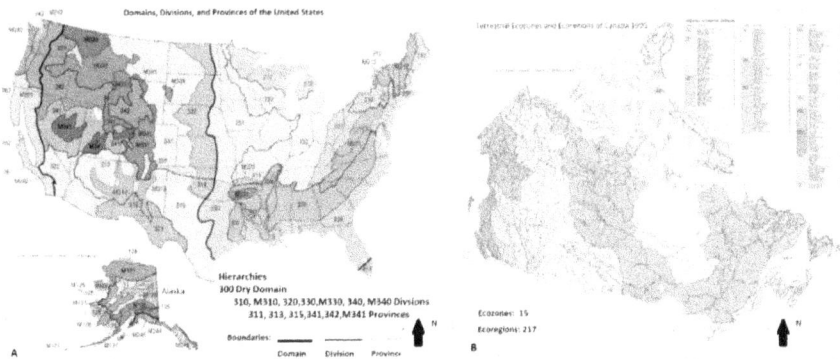

Figure 1.
(A) Upper level ecoregions of the United State. https://www.environment.fhwa.dot.gov/env_topics/ecosystems/ veg_mgmt_rpt/images/vegmgmt_ecoregional_approach_fig 03.png (more detail, refer to the web link). (B) Terrestrial ecozones and ecoregions of Canada. Data source: Environment Canada, Terrestrial Ecozones and Ecoregions of Canada 1995. https://mspace.lib.umanitoba.ca/bitstream/handle/1993/24087/cad_map. jpg?sequence=1&isAllowed=y (more detail, refer to the web link).

2.2 Review of upper level generalizations of Canada

The Ecological Framework from Canada Ecological Stratification Working Group in 1996 defined four upper levels of ecosystems as a nested hierarchy. Definitions and the number of map units for the four levels of generalization are outlined in **Table 1** in Ref. [39] and **Figure 1B** and updated by Statistics Canada in 2018.

In brief, Bailey's 100 Polar Domain only included an area with short summer and low temperature throughout the year, which had been divided into three major Divisions, Icecap Division, Tundra Division, and Subarctic Division, furthermore had been recognized and delimited into 13 Provinces (124,125,126, M121, M125, M126, M127,131,135,139, M131, M135, M139). Bailey also extended Humid Temperate Domain (200) to Canadian Territorial and classified Warn Continental Division (210), Hot Continental Division (220), Marine Division (240), Prairie Division (250), and Dry Domain (300) overlaying with Canada subcontinent. However, the Provinces' descriptions had very little content about Canadian Territory (242, 244,245,251, 331, 332, etc.).

Bailey's 100 Polar Domain overlays the area of Canadian eight Ecozones of Arctic Cordillera (covers Ecoregion 1–7), Northern Arctic (Ecoregion 8–31), Southern Arctic (Ecoregion 32–49), Taiga Plains (Ecoregion 50–67), Taiga Shield (Ecoregion 68–86), Boreal Shield (Ecoregion 87–116), Atlantic Maritime (Ecoregion 117–131), Taiga Cordillera (Ecoregion 165–171) in **Figure 1B**. Furthermore, Bailey's 200 Humid Temperate Domain covers the area of Canadian six Ecozones of Mixedwood Plains (covers Ecoregion132–135), Boreal Plains (Ecoregion 136–155), Prairies (Ecoregion 156–164), Boreal Cordillera (Ecoregion 172–183), Pacific Maritime (Ecoregion 184–197), Montane Cordillera (Ecoregion 198–214). In addition, the Prairies in Canada is extended from 200 Humid Temperate Domain to 300 Dry Domain.

Early pioneering works in North America evolved from forest and climate classifications and were often climate-driven, referred to in [1, 2, 13, 31, 32]. The use of more holistic classifications was recent from 1980′ to 1990′. The holistic approaches were recognized and considered the importance of a broad range of physical and biotic characteristics for identifying ecosystem regionalization and classification. They recognized that ecosystems of any size or level were not always dominated by one particular factor. In describing the ecoregion framework of Canada in [13], Wiken indicated, "The Ecological land classification is a process of delineating and classifying ecologically distinctive areas of the Earth's surface, which can be viewed as a discrete system that has resulted from the mesh and interplay of the geologic,

Ecozones 15	Canada Ecozones on a sub-continental scale is defined and represented an area of the earth's surface of large ecological units classified by interactive and adjusting abiotic and biotic factors. Canada is divided into 15 terrestrial Ecozones.
Ecoprovinces 53	A subdivision of an Ecozone was classified by major assemblages of structural or surface forms, faunal realms, and vegetation, hydrology, soil, and macro climate.
Ecoregions 217	A subdivision of an Ecoprovince was classified by distinctive regional ecological factors, including climate, physiography, vegetation, soil, water, and fauna.
Ecodistricts 1031	A subdivision of an ecoregion was classified by a distinctive assemblages of relief, landforms, geology, soil, vegetation, water bodies and fauna.

*Note: 217 ecoregions and 1031 ecodistricts were updated from 2018 Canada ecological land classification in [38, 39]. E.g. **11.1.165.0858** represented ecozone, ecoprovince, ecoregion and ecodistrict coordinately.*

Table 1.
Upper level ecosystem classification of Canada.

landform, soil, vegetative, climatic, wildlife, water, and human factors.". Therefore, land classification can be applied incrementally on a scale-related basis from site-specific to broad ecosystems.

Because of underlying dynamics of the ecosystems, the multiple patterns of correlation among the biotic, abiotic, and human factors produced the complex; these approaches were apt to produce a converging depiction of regions and significant ecosystem boundary overlapping between Canada and the United States in Refs. [1, 34, 35, 38, 39]. Thus, Canada's continental upper level ecoregion framework defined the ecological Mozaic on a sub-continental scale, representing an area of the Earth's ecological units characterized by interactive and adjusting abiotic and biotic factors. It is not possible to equate Canada and US classification systems directly in [31].

2.3 Implement on lower level ecosystem classification in western Utah of the United States

At Domain, Division, and Province levels, Ecoregions of the United States had been examined by Bailey. The first case study we used for the lower level was accomplished with the upper four levels for the project in a 4.5-million-hectare area centered in western Utah of the United States. National Hierarchy of Ecological Unit (NHEU) had been referenced as the coarsest boundaries in Utah, the United States. This study area was on 300 Dry dominant divisions and had bounders intersecting with 340 Temperate Desert Division and M340 Temperate Desert Regime Mountains Divisions. Three interesting provinces are 342 Intermountain Semi-Desert Province, M341 Nevada-Utah Mountains Semi-desert Coniferous Forest Alpine Province, and 341 Intermountain Semi-Desert and Desert Province. In addition, four sections were intersected in the study area: Bonneville Basin Section, Central Great Basin Section and Northeastern Great Basin Section, and Northwestern Basin and Range Section, shown in **Figure 2**, **Table 2** in [14].

"Bolson" is used as a term in the lower level of ecosystem classification, described the terrain, having entire area from surrounding mountains to mountain slopes, reduced with distance from ridgelines, to the centre of either a river valley or terminal lake basins, or reaching nearly all the study area. DEM data (30 m) was used in the model (**Figure 3A and B**) and generated 60 bolson segments.

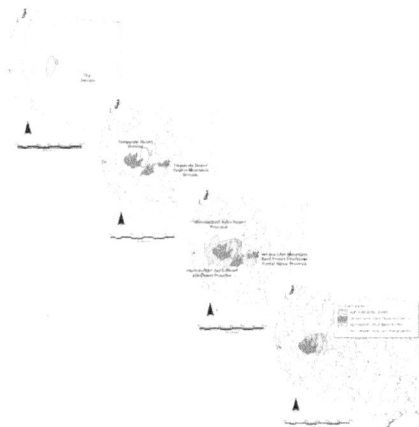

Figure 2.
Upper four levels of ECLs overlaid and intersected in the study area.

Level	ECOMAP name	Example name	Main environmental characters	Scales
1	Domain	300 Dry	Climate/ Köppen Bsk	Ecoregion
2	Division	340 Dry Temperate	Climate	Ecoregion
3	Province	342 Intermountain Semi-Desert	Climate	Ecoregion
4	Section	Central Great Basin	Topography/Terrain	Segment
5	Subsection	Erosional Landscape	Intermediate Scale Terrain Segment	Landscape Mosaic
6	Landtype Association	Hard Erosional Landscape	Macroterrain Units,	Landscape Mosaic
7	Landtype	Eolian Sediments	Mesottrain Units	Landscape Mosaic
8	Landtype Phase	Sedimentary (ridge, slope etc)	Microterrain Units	Zone/Subzone
9	Ecological Site	Desert gravelly Loam	Objectively Defined Land Unit/ Management	Site
10	Vegetation Stand	Sagebrush	Homogeneous Vegetation	Stand

Table 2.
Summaries of the implemented ecosystem classification in western Utah.

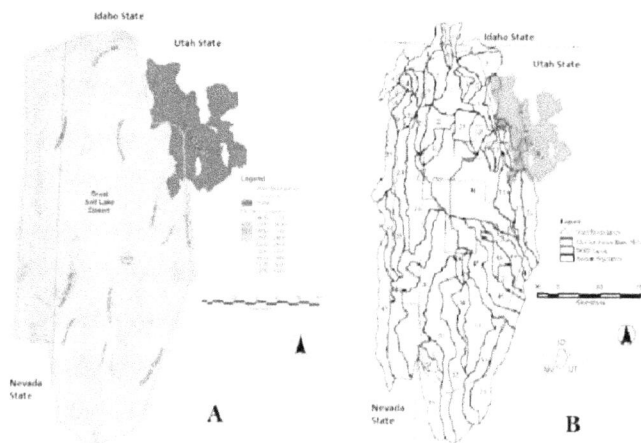

Figure 3.
(A) DEM landscape layout of study area. (B) The 60 bolson segments of the subsection.

2.3.1 Macrotterain units

In the study area, the 60 bolson segments were subdivided into different mac-roterrain units. The algorithm to determine macroterrain units employed elevation and relative change in apparent elevation (slope) from adjacent 30 m DEM cells. It had classified the cells as upslope of equal or higher slope position. Thus, most "mixed" macroterrain unit cells will have "erosional" cells upslope and "deposi-tional" cells downslope depended on their positions. This principle of "superposi-tion" was enforced by the application of the macroterrain class using watershed functions.

2.3.2 Mestoerrain units

With available data of geologic formation or sediments at 1:50,000 scale, the computer algorism was used to identify and delineate the polygons with name attributes for example the metamorphic or moderately hard sedimentary rock, basalt, alluvium, and eolian sediments. By a rationale based on probability, the exposed bedrock units were identified by steeper slope classes, and the presence of rock outcrop as the mapping units.

2.3.3 Microterrain units

The mesoterrain units were divided into subdivisions called microterrain units. Microterrain units were further nested subdivisions of mesoterrain units, which were based mainly on landforms for the erosion-dominated surfaces and landforms plus soils condition. The protocols repeatedly identified landscape units. And two additional levels below the 8th level (NHEU) were added. The 9th level of Ecological sites (ESs) was designed and implemented by using important data on ESs, nested to ECOMAP; the 10th and finest-grain level of vegetation stands were subdivisions of individual polygons of ESs based on differences in disturbance histories (fire, grazing, and human activities) (**Table 2**). The vegetation stands were studied and described by vegetation characteristics, representing fine-scale variations in regional climate, site-specific moisture, nutrient regimes, and disturbance histories (**Figure 4A** and **B**).

2.4 Implement of lower level terrestrial ecoregion classification in Yukon territory of Canada

The major Canadian publications about territorial ecosystem classification or ecoregion classification were designed and generalized as a hierarchical, nested framework with systematic, nested hierarchical layers in the upper four layers (**Table 1**) in [38, 39].

In second case analysis, we validated the Environment Yukon's data and documental report [40–43] with our field observation. The territory of Yukon

Figure 4.
(A) Flow diagram of ecosystem classification of land from bolson segments to vegetation stands. (B) Map of the ecological sites in sampling area.

is approximately 483,450 km2, about 2.2 times that of Utah State in the US, and intersects with Southern Arctic, Taiga Plain, Taiga Cordillera, Boreal Cordillera, and Pacific Maritime Ecozone. Yukon's 23 Ecoregions **of 32, 51, 53, 66, 166–182,184** were described and reported (**Figure 5A**) in [40]. The Yukon Ecosystem and Landscape Classification Framework in [43] provided a classifying tool and method for mapping and implementing ecosystem classification under the Canada Ecozones and Ecoregions.

The research and field work focused on displaying and describing bioclimate features such as the horizontal distribution from south to north and vertical distribution from lower to high (**Figure 5B**). The study was characterized the broad areas influenced by similar climates into a hierarchy of bioclimate zone to lower level classification. Thus, Boreal Low (BOL), Boreal High (BOH), Subalpine (SUB), Taiga Wooded (TAW), Taiga Shrub (TAS), Tundra (TUN), Alpine (ALP) were identified as Bioclimate Zones. The broad ecosystem types by slope position and the phases by plant community dominant species were identified in the nested multiple layers and simplified in **Table 3** in Refs. [41–43]. Field survey and road investigation were carried out at the eleven observation points in 2021 summer (**Figure 5B**). The broad ecosystem types were classified by relative moisture regime as dry, moist, and wet, which can be functionally represented and retrieved the relationship by the generalized the Edatopic Grid as **Figure 6**, and using indexes of Hydrodynamic, aquatic and actual moisture, PH, similarly to it in report [43].

A DEM is a derivative product of the CanVec topographic data set. In Yukon, DEM is available for the entire territory. The generalized GIS model in Keno town area was established to generalize the lower level's bioclimate board ecosystem

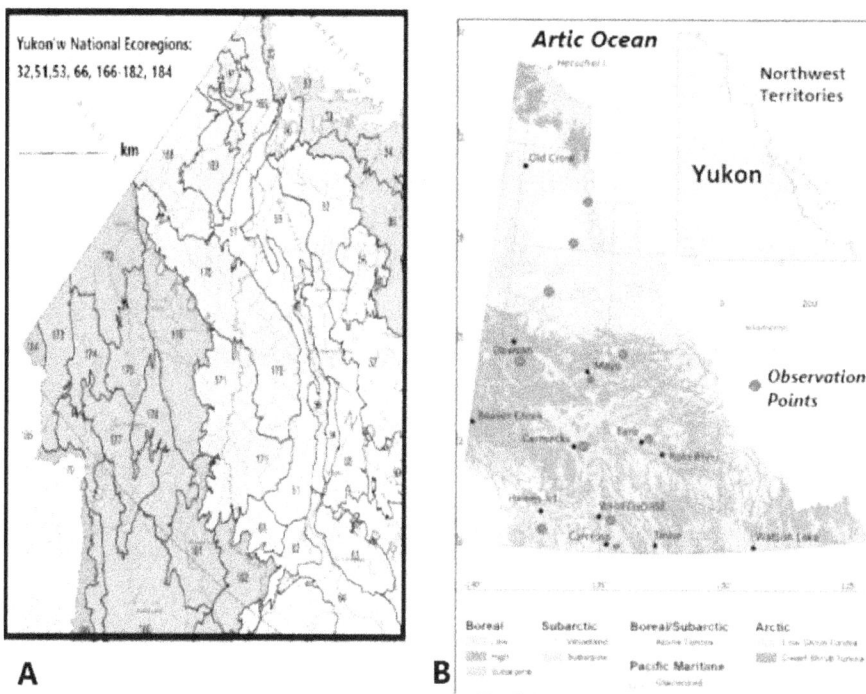

Figure 5.
(A) Yukon ecozones and ecoregions. Data source from Ecological Stratification Working Group and Smith et al. editors [38, 40]. (B) Yukon bioclimate zones, red dot – observation points. Background source from Environment Yukon [43].

Level	Yukon nested ECLs	Classification I	II	III	Equivelent to
1	100 Domain	Domain			Bailey's Top Level
2	12 Ecozone	Boreal Cordilera			Canada's Top level
3	12.2 Ecoprovince	Northern Boreal Cordilera			Bioclimatic Zone
4	12.2.176 Ecoregion	Yukon Plateau-North			Bioclimatic Subzone
5	12.2.176.0898 Ecodistrict	Elsa			Canada ECL's unit
6	Board Ecosystem	H. Wetland	B. Ridge	D. Plains	Bioclimatic/ Slope Position
7	Board Ecosystem Phase	Shrub and salix grasses	Herb	White Spruce	Bioclimatic/ Plants
8	Ecological site/Ecosite	Lodgepole Pine Spruce-Grass- Lichen	Ledium / Salix	Mixedwood/ Boardleaf Forest	Objective or Bioclimatic

Note: bioclimatic Zone: TAW- Taiga Wooded, BOL-Boreal Low, BOH-Boreal High, SUB-Subalpine, TUN- Tundra, ALP- Alpine.
Bioclimatic subzones: Yukon Plateau North, Eagle Plains, North Ogilvie Mountains etc.
Canada ecodistrit can be searched and viewed https://databasin.org/maps/new/#datasets=8dca767690af48e6ae5558 1b34612a19d

Table 3.
Yukon's board ecosystem classification and nested lower levels' ECL.

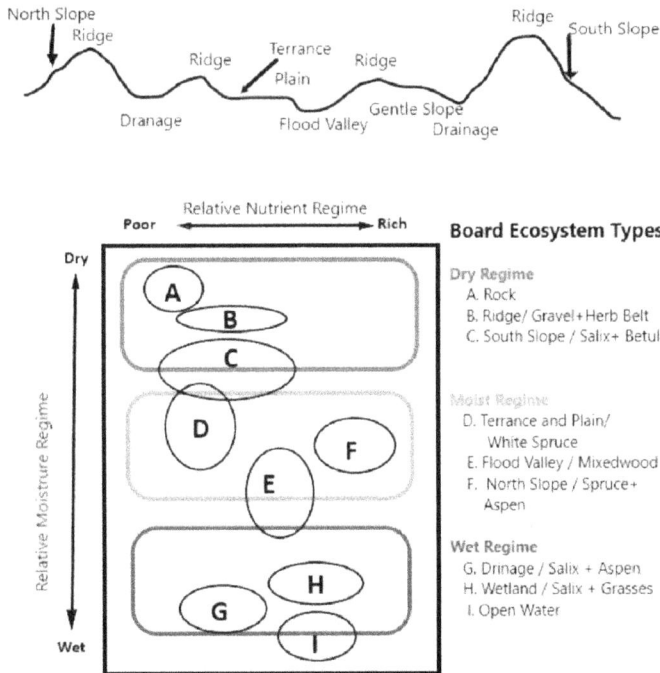

Figure 6.
Broad ecosystem gernerated with edaptopic grid scheme and slope position. The board ecosystem types can be identfied in a lanform position.

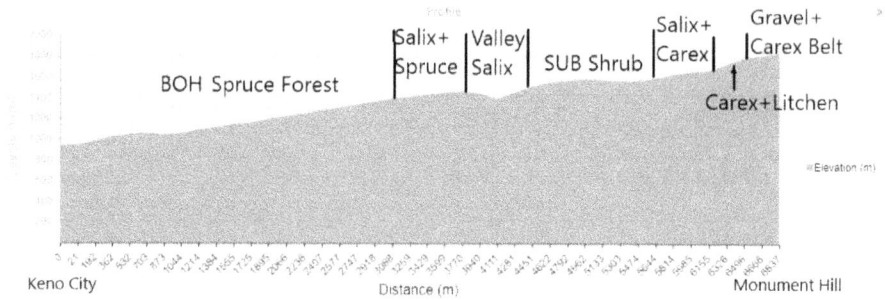

Figure 7.
Vegetation distribution along Keno Hill slope, Yukon.

classification. Predictive ecosystem mapping relayed on digital elevation models (DEM) to represent landform slope and aspect conditions. These conditions provided and informed soil moisture, a primary determinant of ecosystem pattern. A demonstration was the slope survey completed near Keno city up to Monument hill (**Figure 7**). Subalpine shrub appeared above elevation 1530 m, and Salix + Carex shrub grasses from 1600 m to 1730 m. Homogenous Carex + Litchen alpine vegetation located at 1780 m become biological indicator where was near the ice valley or cold environment. Gravels + Carex + gravels belt located at 1825 m indicated that the seasonal frozen condition was occurred constantly.

3. Discussion

By analyzing the upper level of ECLs in the United States and Canada, we realized that the ecosystem classification of land was a special methodology to explore and classify the ecoregions in the different countries. Bailey classified upper-level Ecosystem Classification of Land (Domain, Division, and Province), in which Domain was based on Köppen climate system classification [1–3]. Bailey, in Ref. [34], indicated that the differences in the climatic regime distinguish the natural ecosystems. The principle is that climate, as a source of energy and moisture, acts as the primary control for the ecosystem. Whether or not using Bailey's Domain as the top level of Canada's territorial Domain remained a further comparison between the United States and Canada. At least, the upper four levels' ecosystem classification and detail descriptions of Canada (see **Table 1**) would be the best fulfillment and data source. Technically, the vector and raster data can be retrieved and integrated into GIS software [14, 44–46].

The Ecological Framework of Canada in Refs. [37–39] used different classification schemes and presented the upper four levels of ecosystem classification with features of hierarchy structure in a subcontinent scale. Canada's top-level fifteen Ecozones have overlaid and intersected with Bailey's 100 Polar Domain, 200 Humid Temperate Domain, and 300 Dry Domain. For instance, Bailey's 100 Polar Domain overlays the area of Canadian eight Ecozones, Bailey's 200 Humid Temperate Domain covers the area of Canadian six Ecozones. In addition, the Prairies in Canada is extended from 200 Humid Temperate Domain to 300 Dry Domain in the US.

ECOMAP defined by the National Hierarchy of Ecological Unit (NHEU), had presented the "top-down" approach of Ecosystem Classification of Land in the United States. Western Utah's project had proved that it was a cost matter through a complete ECL's field survey. Another consequence of the strictly top-down nested hierarchical design of ECOMAP is that progressively smaller and unique polygons are created for each level. In other words, the ECOMAP process applied so far prevents one from easily relating features at one location to those within other land-form units or bolson segments. Thus, ECOMAP is a top-down regionalization with hierarchically nested features for an explicitly geographic area. At the same time, these futures allow the ecosystem classification units to be used for various needs, from local to national. These features in the NHEU are the perimeters of outer polygons created at lower levels have to be vertically integrated with the delineation of polygons occurring at upper levels.

The limitation is for this "top-down" process; if the lowest levels are produced independently from higher levels, we still cannot answer whether the similarity of the same label polygon or unit is the same until a field survey is conducted or references available.

Much information for local managers and management companies, not all information very useful for Ecological land of classification. We did not expect any ecological research had funding to complete for mapping as to details. The project in a dry domain area with a 10 level classification would be more theoretical than practical management.

While network linked rather than nested hierarchically could be employed, we propose a simpler, more straightforward solution. Our actions were carried out a complete hierarchical land classification from a top-down approach. Ideally, we treated the ecosystem like an "organism" and separated it into components, following a top-down nested hierarchy to its finest subdivisions, and countered in common sense and practicality. Thus, a terrestrial ecosystem is considered as a volume of earth space with organic contents. We separated it from its neighbors by reasonable divisions by the empirical observation and knowledge in climatology, geography, ecology, soil, and physiography in [47–51].

While it is recognized that the National Ecological Framework with the terrestrial ecoregions in **Table 1** is a referential part of the Yukon ELC Framework, maintaining these layers for Yukon as attributive layers and data in the GIS model that is recommended in [40–43]. Specially, using 100 Domain as a top level ELC. Canada's Ecozone was considered as second level ELC. Canada's Ecoprovince in Yukon Territory was equivalent to the Bioclimate Zone, and Ecoregion was equivalent to the Bioclimate Subzones. Canada's Ecodistrict was established and can be used as identical fifth ELC layer. The sixth and seventh ELCs were related to Bioclimatic Board Ecosystem in terms of slope position and plant population important index. Canada's eight ELC was objectively defined Ecological Site or bioclimatic Ecosite. Thus, we established a complete ELC in Yukon Territory (**Table 3**).

The management approach and applications for the broad ecosystem classification and mapping are listed in **Table 4**.

Mapping level and scales	Applications	Context
Bioclimate (1:100,000 to 1,000,000)	Climate Change Studies	Plant species shifting and community succession
Board Ecosystem 1:50,000 1:250,000	Regional land use planning	Land use changes and management policy
Local Ecosystems 1:10,000 to 1:50,000	Environmental Impact assessments	Land Degradation, recovery and restoration
Varies	Ecosystem Services	Ecosystem Assessment, Supporting, provisional, regulating and cultural services

Table 4.
Broad ecosystem classification mapping and applications.

Practically, the lower level cases of Canada territorial Ecosystem Classification had preferred more practice and objective. The researchers can use GIS technology and Spatial Analysis Modeling to efficiently produce the different maps for the landowner, management companies, and government agencies. In addition, plant ecologists had sophistical experiences in [18, 30, 33, 44, 52–57] to develop the vegetation classification and ecoregion map with a nested structure using biogeoclimatic principles. The map products were delivered by the scaled-based ecosystem classification and represented them with a high relation among the long-term climate condition, climax vegetation, and dominant plant species.

In addition to Bioclimatic Board Ecosystem Classification, Biogeoclimatic Ecosystem Classification (BEC) approach was often demonstrated as a quick approach and identified as an ecological framework for vegetation classification, mapping, and monitoring vegetation dynamics in [33, 44, 53–55, 58]. BEC approach has been used in many provinces in Canada, and the association-based ecological units of BEC are the fundamental units, for example, that the boreal vegetation association was integrated for its boundary justification. Also, the BEC approach delineated ecologically equivalent climatic regions and displayed the site conditions in the Edatopic Grid with a relationship between soil nutrient regime and soil moisture regime in [53, 54].

Ecologists studied different computational models in ecological classification such as LeNet, AlexNet, VGG models, residual neural network, and inception models in Refs. [16, 17, 24, 28]. The biggest challenge was faced in the need for an extensive training dataset to achieve high accuracy. Examples trained algorithms and the machine can only detect what criteria have been previously shown and selected. Deep learning, or machine learning algorithms, was going on method for analyzing nonlinear data with complex interactions. Moreover, they can achieve remarkable accuracy for identification and classification tasks. As a result, achieving proper ecological predictions is more feasible now. Increasing data availability is highly related to using GIS, remote sensing, and international research networks in Refs. [45, 46, 56, 57]. Furthermore, a fundamental change in research culture is towards making ecological data open access publically. All of these developments are important factors behind deep learning and development in ecology.

With further understanding, the ecosystem classification approaches and ecological modeling experiences in [14, 44, 46, 56, 57, 59] and objectively defined

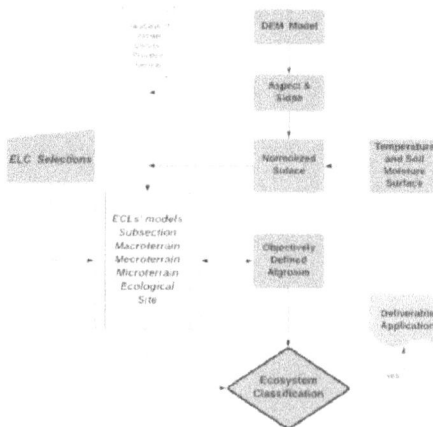

Figure 8.
Objectively defined ecosystem classification.

ecosystem classification can be integrated by using a computer algorithm to develop efficient tools and affordable applications (**Figure 8**) without losing hierarchical structure feature in [30]. The ECL menu had input data function by getting upper-level Domain, Division, Province, and Section digital format data, and carried out a deliverable application associated with a scaled lower level ECLs. The objective analysis generated internal function outputs and combined them in the Deep Learning Algorism. The slope model, landform model, was running based on objective needs; vegetation, soil, and geology data could be considered attribute data sources depending on the study area.

We did not discuss landscape-scale changes and boundary issues that influenced ecosystem classification, which authors already presented in Refs. [1, 2, 11, 15, 25, 31, 48, 49]. Second study case demonstrated that a full ECL generally included three components: Bailey's upper level ECL, Broad Ecosystem classification, and bottom level Ecological site. With assessment, justification, and testing, we completed a full Ecosystem Classification in a Yukon ecoregion.

Why do we use western Utah's ECL to compare with Yukon's? The direct reason is that these two ecoregions had fewer human activities and had more broad original nature ecosystems in North America. In the meantime, the climate conditions are between a Dry Domain and a Polar Domain in these two ecoregions. Our study cases led the research and study with a complete ELC in Bailey's 300 Dry Domain and 100 Polar Domain.

4. Conclusions

Canada's continental upper level ecoregion framework defined the ecological Mozaic on a sub-continental scale, representing an area of the earth's ecological units characterized by interactive and adjusting abiotic and biotic factors. Therefore, using Bailey's Domain as the top level of Canada's territorial ecoregion was recommended. Similarly, many users suggested that they examined the popularity and characteristics in a study area linked to the continental and global scales in [1, 8, 59–62] whenever necessary and integrated to delineate and identify the regional ecosystem. Ecological regionalization is an abstraction from global to a local site-level, contributing to understanding nature and providing differentiated guidance to sustainable environmental management. It recommended that using the global ecoregion scheme offers the guidelines for biodiversity conservation, but it still faces obstacles in improving ecosystem services and substantial uses. We had reviewed and analyzed the regionalization process, implements in two ecoregions, and some practices. With the critical consideration of ecosystem services, global environmental change and human activities should be followed in functionalized ecological regionalization. Ecosystem regionalization is a scale-based approach to classifying land surfaces, combined with regional and continental data. We should have understood more about taking geology, landform, soils, vegetation, and climate into account to classify the regionalization in different scales and ecosystem levels for a global-wide scheme when the ecosystem studies and services have grown in the research, publication and practice.

Acknowledgements

Correspondence author collaborated in USU's ECL project with Prof. Neil West (Referred West et al., 2005), and conducted the Yukon Ecosystem Classification

Project. The final study was supported by Instant Calling Spatial Arch Lab, Burnaby, BC, Canada. Thanks to Prof. Neil West for his past advice. Thanks to Simon Fraser University Library funds for eligible open access publication.

Authors' contributions

Authors have contributed a lot to this manuscript and approved the final manuscript.

Funding

Final stage's research fund was supported by Instant Calling Spatial Arch lab, Burnaby, B.C. Canada, and Simon Fraser University Library Open Access Fund.

Conflict of interest

Authors declare that there are no competing interests.

Author details

YanQing Zhang[1*] and Neil E. West[2]

1 School of Computing Science, Department of Geography, Simon Fraser University, Burnaby, Canada

2 Department of Range Science, Utah State University, Logan, UT, USA

*Address all correspondence to: yqz@sfu.ca

IntechOpen

References

[1] Bailey R. Description of the ecoregions of the United State. 2nd ed. Rev. and expanded (1st ed. 1980). Misc. Publ.No. 1391(Rev). Washington DC USDA Forest Service. 1995.

[2] Bailey R. Ecosystem Geography. Springer-Verlag. New York. 1996.

[3] Bailey R. Identifying Ecoregion Boundaries. Environmental Management. 1983; 34 (1): 14-26, DOI: 10.1007/s00267-003-0163-6

[4] Walfram S. Science is a computer program, If complexity arises from simple rules, should we rethink how to do science? Wolfram media. 2002; 1: 199.

[5] Clark J, Carpenter S, Barber M, Collins S, Dobson A, Foley J A. et al. Ecological forecasts: an emerging imperative. Science 2001; 293:657-670. doi: 10.1126/science.293.5530.657

[6] Wallace K. Classification of Ecosystem Services: Problems and solutions. Biological Conservation 2007; 139: 235-246.

[7] O'Neill R, DeAngells D, Waide J, Allen T. A Hierarchical Concept of Ecosystems. 1986. Princeton University Press, New Jersey.

[8] Bailey R, Hogg H. A world ecoregions map for resource reporting. Environmental Conversation 1986; 12:195-202. https://doi.org/10.1017/S0376892900036237.

[9] Albert D, Lapin M, Pearsall D. Knowing the territory: Landscape Ecosystem and Mapping. The Michigan Botanist. 2015; 54: 34-41. http://hdl.handle.net/2027/spo.0497763.0054.106.

[10] Hutchinson, M, McIntyre S, Hobbs R, Stein J, Garnett S. and Kinloch J. Integrating a global agro-climatic classification with bioregional boundaries in Australia. Global Ecology and Biogeography. 2005; 14: 197-212. https://doi.org/10.1111/j.1466-822X.2005.00154.x.

[11] Klijn F, de Haes H. A hierarchical approach to ecosystems and its implications for ecological land classification. Landscape Ecol 1994; 9: 89-104. https://doi.org/10.1007/BF00124376

[12] Creque J, Bassett S, & West N. Viewpoint: delineating ecological sites. Journal of Rang Management. 1999: 52; 546-549. http://hdl.handle.net/10150/644040.

[13] Wiken, E. (compiler). Terrestrial ecozones of Canada. Ecological Land Classification Series No. 19. 1986. Hull, QC: Environment Canada.

[14] West N, Dougher F, Manis G, Ramsey R. A comprehensive ecological land classification for Utah's West Desert. Western North American Naturalist. 2005; **65**(3): 281-309. https://works.bepress.com/neil_west/24/

[15] Bourgeron P, Humphries H, and Jensen M. Ecosystem Characterization and Ecological Assessments in M. E. Jensen et al. (eds.), A Guidebook for Integrated Ecological Assessments © Springer Science+Business Media New York; 2001. https://doi.org/10.1007/978-1-4419-8620-7_4

[16] De'ath D, & Fabricius K. Classification and Regression Trees: A Powerful Yet Simple Technique for Ecological Data Analysis. Ecology. 2000; 8(11): 3178-3192. https://doi.org/10.1890/0012-9658(2000)081[3178,CARTAP]2.0.CO;2.

[17] Allen C, Angeler D, Garmestani S, Gunderson H, Holling S. Panarchy: theory and application. Ecosystems.

2014; 17: 578-589. DOI: 10.1007/s10021-031-9744-2.

[18] Hou XY. Editor. Vegetation of China with reference to its geographic distribution, Annals Missouri Bot. Gard. 1983; **70**: 509-548.

[19] Zhang YQ, & Zhou XM. The quantitative classification and ordination of Haibei alpine meadow. Acta Phytoecological ET Geobotanica Sinica. 1992; **16**(1): 36-42. https://www.plant-ecology.com/EN/Y1992/V16/I1/36.

[20] Cleland D, Avers P, McNab W, Jensen M, Bailey R, King T, Russell W. National Hierarchical Framework of Ecological Units. Published in, Boyce, M. S.; Haney, A., ed. Ecosystem Management Applications for Sustainable Forest and Wildlife Resources. Yale University Press, New Haven, 1997; 181-200.

[21] Zheng D. A Study on the Eco-geographic Regional System of China, FAO FRA2000 Global Ecological Zoning Workshop, Cambridge, UK. 28-30 July 1999. P 12. http://www.fao.org/3/ae344e/ae344e09.htm.

[22] Wu SH, Yang QY, Zheng D. Comparative Study on Eco-geographic Regional Systems between China and USA[J]. Acta Geographica Sinica. 2003; **58**(5): 686-694. http://www.geog.com.cn/EN/Y2003/V58/I5/686.

[23] Wu SH, Yang QY, Zheng D. Delineation of eco-geographic Regional Systems of China. Journal of Geographical Science. 2003; **13**(3): 209-315. http://www.geogsci.com/EN/Y2003/V13/I3/309.

[24] Brodrick P, Davies A, Asner G. Uncovering Ecological Patterns with Convolutional Neural Networks. Trends in Ecology & Evolution. 2019; **34**: 1-12. Doi: 10.1016/j.tree.2019.03.006

[25] Bourgeron P, Humphries H, Jensen M. Elements of Ecological Land Classifications for Ecological Assessments. A Guidebook for Integrated Ecological Assessments 2001; 22: 321- 337 Springer New York. https://doi.org/10.1007/978-1-4419-8620-7_23

[26] ECOMAP. National hierarchial framework of ecological units. U.S. Department of Agriculture, Forest Service, Washington, DC. 1993. https://www.researchgate.net/publication/237419014_National_hierarchical_framework_of_ecological_units.

[27] Bailey R, Jensen M, Cleland D, and Bourgeron P. Design and use of ecological mapping units, In: M.E.Jensen and P.S.Bourgeron, (eds.), *Ecosystem Management, Volume 2: Principles and Applications*, Gen. Tech. Report No. PNW-GTR-318, US Dep. Agric., For. Serv., Pacific Northwest Research Station, Portland, Oregon. 1994. p. 95-106.

[28] Olden J, Lawler J, Poff N. Machine learning methods without tears: a primer for ecologists. Q. Rev. Biol. 2008; 83: 171-193. doi: 10.1086/587826

[29] Hornsmann I, Pesch R, Schmidt G. and Schröder W. Conference paper: Calculation of an Ecological Land Classification of Europe (ELCE) and its application for optimising environmental monitoring networks. 2008, 1-12. https://www.researchgate.net/publication/275039549

[30] Zhang YQ. A Hierarchical Analysis On Ecosystem Classification with implementing in Two Continental Recregions. Environmental Systems Research. 2021;**10**:39. DOI: 10.1186/s40068-021-00243-3

[31] Bailey R, Zoltai S, Wiken E. Ecological regionalization in Canada and the United States. Geoforum. 1985; 16(3):265-275. https://doi.org/10.1016/0016-7185(85)90034-X

[32] Bailey R. Design of Ecological Networks for Monitoring Global Change. Environmental Conservation. 1991;18(2): 173-175. doi: 10.1017/S0376892900021780.

[33] Baldwin K, Chapman K, Meidinger D et al. The Canadian National Vegetation Classification: Principles, Methods and Status. Natural Resources Canada, Canadian Forest Service Information Report GLC-X-23. 2019; 1-163. DOI: 10.13140/RG.2.2.25385.95847.

[34] Bailey R. Suggested hierarchy of criteria for multi-scale ecosystem mapping, Landscape and Urban Planning. 1987; 14:313-319. ISSN 0169-2046. https://doi.org/10.1016/0169-2046(87)90042-9.

[35] Bailey R. Ecosystem Geography: From Ecoregions to Sites. 1996. Springer-Verlag. New York.

[36] Commission for Environmental Cooperation. Ecological Regions of North America: Towards a Common Perspective. Montreal, Quebec. 1997.71 pp. Map at scale 1: 12.5 million. Ecomaps. ISBN: 2-922305-18-X.

[37] Lands Directorate. Terrestrial Ecozones Of Canada, Ecological Land Classification No. 19. 1986. p. 26.

[38] Ecological Stratification Working Group. A National Ecological Framework for Canada. Agriculture and Agri-Food Canada, Research Branch, Centre for Land and Biological Resources Research and Environment Canada, State of Environment Directorate, Ottawa/Hull. 1996. p125. And Map at scale 1:7.5 million. from /cansis/publications/ecostrat/index.html

[39] Statistics Canada. Ecological Land Classification. 2018. Catalogue no. 12-607-X.

[40] Smith C, Meikle J, and Roots C. (editors). Ecoregions of the Yukon Territory: Biophysical properties of Yukon landscapes. Agriculture and Agri-Food Canada, PARC Technical Bulletin No. 04-01, Summerland, British Columbia. 2004. p 197-206.

[41] Grods J, Francis S, Meikle J. and Lapointe S. *Regional Ecosystems of West-Central Yukon, Part 1: Ecosystem descriptions.* Report prepared for Environment, Government of Yukon by Makonis Consulting Ltd. and Associates, West Kelowna, BC. 2012.

[42] Grods J, Francis S, Meikle J. and Lapointe S. Regional Ecosystems of West-Central Yukon, Part 2: Methods, input data assessment and results. Report prepared for Environment, Government of Yukon by Makonis Consulting Ltd. and Associates, West Kelowna, BC. 2012.

[43] Environment Yukon. Flynn N. and Francis S, editors.Yukon Ecological and landscape Classification and Mapping Guidelines. Version 1.0. Whitehorse (YT): Department of Environment, Government of Yukon. 2016.

[44] MacKenzie W. & Klassen R. VPro 13: Software for management of ecosystem data and classification. 2009. Version 6.0. BC. Min. For. and Range, Research Branch, Victoria BC. URL: https://www.for.gov.bc.ca/hre/becweb/resources/ software/vpro/index.html

[45] Iwao K, Nishida K. et al, Creation of New Global Land Cover Map with Map Integration. Journal of Geographic Information System.2011; **3**: 160-165. DOI: 10.4236/jgis.2011.32013.

[46] Zhang L, Xia G, Wu T, Lin L, Tai X. Deep Learning for Remote Sensing Image Understanding. *Journal of Sensors*, vol. 2016; Article ID 7954154, 2 pages. https://doi.org/10.1155/2016/7954154.

[47] Rowe J. The level of integration concept and ecology. Ecology. 1961;

42:420-427. https://doi.
org/10.2307/1932098

[48] Rowe J. Land classification and ecosystem classification. Environ Monit Assess. 1996; 39: 11-20. https://doi. org/10.1007/BF00396131R

[49] Rowe J. The integration of ecological studies. Functional Ecology. 1992; 6: 115-119.

[50] Rowe J. Viewpoint. Biological fallacy: life equals organisms. BioScience. 1992; 42(6): 394.

[51] Rowe J. and Barnes B. Geo-ecosystems and bio-ecosystems. Bull. Ecol. Soc. Amer. 1994; 75(1): 40-41.

[52] Zhou XM, Wang Zh & Du Q (1987) Qinghai Vegetation. 1987. Qinghai People Press in Chinese.

[53] MacKenzie W, Medidinger D. The Biogeoclimatic Ecosystem Classification Approach: an ecological framework for vegetation classification. Phytocoenologia. 2017; 1-11. DOI: 10.1127/phyto/2017/0160.

[54] McLennan D, MacKenzie W, Meidinger D, Wagner J, and Arko C. A Standardized Ecosystem Classification for the Coordination and Design of Long-term Terrestrial Ecosystem Monitoring in Arctic-Subarctic Biomes. Arctic. 2018; 71: SUPPL. 1, 1-15. DOI: https://doi.org/10.14430/arctic4621.

[55] DeLong S, Griesbauer H, Mackenzie W, and Foord V. Corroboration of Biogeoclimatic Ecosystem Classification climate zonation by spatially modelled climate data. BC Journal of Ecosystems and Management. 2010; 10(3):49-64. www. forrex.org/publications/jem/ISS52/ vol10_no3_art7.pdf

[56] Zhang YQ, Peterman M, Aun D, & Zhang YM. Cellular Automata:

Simulating Alpine Tundra Vegetation Dynamics in Response to Global Warming. Arctic, Antarctic, and Alpine Research. 2008; 40(1): 256-263. http:// www.jstor.org/stable/20181786.

[57] Zhang YQ, Sun, MH, Welker J. Simulating Alpine Tundra Vegetation Dynamics in Response to Global Warming in China. Chapter in book: Global Warming. 2010. Chapter 11. p. 221-250. DOI: 10.5772/10291.

[58] Kremer M, Lewkowicz A, Bonnaventure P, & Sawada M. Utility of Classification and Regression Tree Analyses and Vegetation in Mountain Permafrost Models, Yukon, Canada. Permafrost and Periglacial Processes, 2011; 22(2), 163-178. https://doi. org/10.1002/ppp.719

[59] Liu YX, Fu BJ, Wang Sh, Zhao WW. Global ecological regionalization: from biogeography to ecosystem services, Current Opinion in Environmental Sustainability. 2018; 33: 1-8. https://doi. org/10.1016/j.cosust.2018.02.002.

[60] George E, Philip L, Mladenoff D, White M, Crow T. A Quantitative Approach to Developing Regional Ecosystem Classifications, Ecological Application. 1996; 6(2):608-618. https://doi.org/10.2307/2269395.

[61] Mueller M, Pander J & Geist J. A new tool for assessment and monitoring of community and ecosystem change based on multivariate abundance data integration from different taxonomic groups. Environ Syst Res. 2014; 3, 12. https://doi.org/10.1186/2193-2697-3-12.

[62] Mengist W, Soromessa T. Assessment of forest ecosystem service research trends and methodological approaches at a global level: a meta-analysis. Environ Syst Res. 2019; 8: 22. https://doi.org/10.1186/ s40068-019-0150-4.

Chapter 8

Spinless Forage Cactus: The Queen of Forage Crops in Semi Arid Regions

Marcelo de Andrade Ferreira, Luciano Patto Novaes,
Ana María Herrera Ângulo
and Michelle Christina Bernardo de Siqueira

Abstract

Forage cactus is a perennial crop, which has been widely exploited for feeding ruminants in the semiarid region of different countries around the world. The main objective of this chapter is to describe the use and importance of spineless cactus as forage, desertification mitigation, source of water for animals and a source of income for producers in semiarid regions. The main species explored in Brazil are *Opuntia* spp. and *Nopalea* spp., due to characteristics such as resistance to pests, productivity, water-use efficiency and demand for soil fertility. The productivity of the species in a region will depend on its morphological characteristics, plant spacing, planting systems and its capacity to adapt to climatic and soil conditions. In other parts of the world, cactus species are the most cosmopolitan and destructive among invasive plants. However, the use of spineless forage cactus in areas where it can develop normally and may become the basis for ruminants' feed would increase the support capacity production systems. Thus, specifically for Brazil's semiarid region these species can make the difference as forage for animal feeding, cultivated as monoculture or intercropped, for soil conservation and desertification mitigation, source of water for animals, preservation of the Caatinga biome and be a potential source of income for producers if cultivated as vegetable for nutritional properties and medicinal derivative of fruits and cladodes for exports.

Keywords: livestock, smallholder, sustainability, energy

1. Introduction

Spineless Forage Cactus is no doubt a magic forage plant having potential to serve as a source of water bank and forage for animals under extreme environment, but it does not fall under the scope of book Grasses and grasslands: New perspectives. Due to its resistance to drought and high efficiency in the use of rainwater, the planting and use of Spineless Forage Cactus is neglected in semi-arid regions, which is a mistake. In these regions and suitable climatic conditions, it is an unbeatable crop in terms of productivity and quality as an energy food, which is why it has the power to be called the Queen of Forages in the Semiarid Region.

Scientific production around this forage crop dates back to the 1980s, with increasing interest in recent years, mainly in countries such as Mexico, Tunisia, the United States, Argentina, India and Brazil. Recently, it highlights scientific production related with crop productivity as a monoculture or intercropped, mineralization dynamics of differents sources of organic fertilizers, irrigation, its use as a food supplement or ingredient substitute and how ruminants supplement with spineless cactus can reduces drinking water ingestion.

This chapter is intended to describe a brief use and importance of spineless cactus as forage, desertification mitigation, source of water for animals and a source of income for producers in semiarid regions. As methodology, published papers on planting methods and cultural treatments were researched, aiming at the knowledge of those that allow greater productivity and also articles related to nutritional value that would allow its recommendation as the main alternative as a source of energy for ruminants in semiarid regions. Finally, simulations were carried out in order to demonstrate that the use of forage catus could help in environmental conservation. Papers are located from physical and virtual libraries.

2. Stand, productivity, and spacing

Forage cactos as *Opuntia* and *Nopalea* are a perennial crop, developed in several semiarid regions [1]. During periods of drought, it is used as forage in countries such as United States, Mexico, South Africa, Australia, Tunisia, Egypt, and Brazil [1–4]. In Brazil, It was introduced in 1880 and it is considered the main source of feed for herds, mainly in the semiarid region [1, 5, 6]. Its taxonomy is widespread among vascular plants and it is present in many succulent species from semiarid regions [7].

According to the Agricultural Census [8], the production of forage cactus in the semiarid region of Brazil is 3,581,469 tons, with productivity of 24.3 t/ha of dry matter in a harvested area of 147,439 ha. This production is concentrated in the states of Bahia (1,500,359 ton), Paraíba (742,982 ton), Pernambuco (481,932 ton) and Sergipe (431,468 ton).

The main species explored in Brazil are *Opuntia* spp. and *Nopalea* spp. For decades, the varieties of *Opuntia ficus-indica* have been considered among those the best establishment, after introduction into a new area, more resistant to drought or adverse conditions, long shelf life, and most productive [9, 10]. However, they are the most sensitive species to attack by the cochineal insect [*Dactylopius opuntiae* (Cockerell)]. As a result, more resistant varieties are expanding, the clones are IPA Sertânia [IPA; *Nopalea cochenillifera* (L.) Salm-Dyck], Miúda (*N. cochenillifera*), Mexican Elephant Ear [OEM; *Opuntia stricta* (Haw.) Haw.] and African Elephant Ear (OEA; *Opuntia undulata* Griffiths; [10–12]. There are still many plantations with the variety *O. ficus-indica* in Brazil [13]. However, the authors highlight the need to diversify the genetic base, introducing new genotypes, mainly due to the occurrence of the cochineal insect.

In many cases, despite belonging to the same genus, forage cactus species present different responses under different growing conditions. Thus, the productivity of the species in a region will depend on its morphological characteristics [14] and its capacity to adapt to climatic and soil conditions (**Table 1**) [6, 15].

The variety OEM is an imported clone native from Mexico which has been highlighted by its greater tolerance to drought, resistance to *D. opuntiae*, and high productivity [6, 10, 20]. More recently, it has been highlighted by its higher forage productivity, water accumulation, water use efficiency, and carrying capacity [18].

The recommended plant spacing for forage cactus varies according to the production system and the environment, and it can be planted as a single crop or

Clone	Plants/ha	Spacing	Harvest frequency	DMP (t/ha)	Reference
Ipa Sertânia[1]	28,000	1.6 × 0.2 m	2 years	10.7	[6]
Miúda[1]	20,000	1.0 × 0.50 m	2 years	7.35	[16]
Miúda[1]	29,875	1.6 × 0.2 m	2 years	11.5	[6]
OEM[2]	30,938	1.6 × 0.2 m	2 years	15.6	[6]
OEM[2]	33,333	2.2 × 0.2 m	234 days	13.7	[17]
OEM[2]	25,000	1.0 × 0.4 m	330 days	16.4	[18]
Gigante[3]	20,000	1.0 × 0.5	600 days	21.5	[19]
	20,000	3.0 × 1.0 × 0.25m	600 days	14.7	

[1]*Nopalea cochenillifera (L.) Salm-Dyck.*
[2]*Orelha de Elefante Mexicana [Opuntia stricta (Haw.)].*
[3]*Opuntia ficus-indica Mill; DMP: dry matter production.*

Table 1.
Productivity of forage cactus clones under dryland condition.

intercropped with commercial crops [21]. In a single crop, there is greater proximity between plants, especially in double rows, which can favor greater competition for nutrients, damaging growth [19]. However, according to [22] it is possible to obtain greater productivity in dense crops due to the increase in the number of plants per hectare and, consequently, the increase in the cladode area index. However, depending on the genotype-environment combination, there will be a limit where light interception and photosynthetic efficiency can be affected. If mechanization is available, this must also be taken into account when choosing the optimal spacing [21]. Less dense plantings facilitate cultural treatments and reduce the risk of pests such as cochineal insect [22]. According to [23] it is possible to use planting arrangements in triple or quadruple rows that favor the mechanization of the forage cactus *O. ficus-indica* Mill. Although, this can affect sustainability, since the increase in the area covered by plants reduces erosion processes, favoring the maintenance of the most fertile layers in the soil [24]. Some examples of the importance and variability of the productive response to planting spacing and density are highlighted in **Table 1**.

Intercropping planting systems can also affect the productivity and harvest timing of forage cactus [14]. Some of the crops considered in these intercropping systems have been, *Vigna unguiculata* (L.) Walp, *Sorghum bicolor* L. [14, 25], *Spondias* spp. [26], *Leucaena leucocephala* (Lam.) by Wit., and *Gliricidia sepium* (Jacq.) Steud. [27]. Different responses were observed highlighting clone importance. In *O. stricta* (OEM), the cutting season of forage cactus was anticipated (17 months), indicating that competition with sorghum did not reduce its monthly growth rate [14]. While for *Nopalea cochenillifera* (L.) Salm-Dyck., there was no difference in production (20.5–24.5t/ha) concerning the single crop [27]. For all referenced works on cactus intercropped with grasses or legumes, morphophysiological and productive changes were verified in relation to the growth dynamics of both cultures. However, recommendations for resilient production systems can be useful under semiarid conditions.

The consortium of forage catus and the use of appropriate management practices can contribute to improve soil fertility, increase crop productivity and the sustainability of livestock production systems. Northeastern semi-arid region. The introduction of Leucaena (*Leucaena leucocephala* (Lam.)) or Gliricidia (*Gliricidia sepium* (Jacq.) Steud.) intercropped with forage cactus, along with the application of manure, is a relevant alternative for production systems in the semi-arid

Clone	Dry matter production (t/ha)[1]			Plants/ha	Harvesting frequency	Reference
	Basal	Primary	Secondary			
Miúda[2]	11.03	17.5	23.04	50,000	12 months	[32]
Gigante[3]	8.62	14.83	19.64	50,000	12 months (year 1)	[30]
	14.9	22.3	34.7		12 months (year 2)	
Gigante[3]	—	3.9	—	—	12 months	[33]
	—	13.2	—	—	24 months	
OEM[4]	20.9	37.5	33.2	43,478	12 months	[34]

[1]*Preserving corresponding cladode.*
[2]*Nopalea cochenillifera (L.) Salm-Dyck.*
[3]*Opuntia ficus-indica Mill.*
[4]*Orelha de Elefante Mexicana [Opuntia stricta (Haw.) Haw.].*

Table 2.
Forage cactus production under different cutting intensities and harvest time.

region, in order to increase soil organic matter and soil nutrients because deposition of litter with low C: N ratio. Such improvements imply the maintenance of soil fertility, cactus productivity, and the sustainability of these systems. The forage cactus can be intercropped with several crops, whether annual or perennial, such as corn, sorghum, beans, sunflower, pigeon pea, gliricidia among others [28], but researches with forage catus intercropped with other cultures are recent and are not conclusive.

A decrease in dry matter production of 22.7% and 39.2% of forage cactus and sorghum, respectively, when they were cultivated in intercropping [29].

The cutting intensity and harvest management of forage cactus are two other factors that affect crop productivity. The efficiency of plants in converting light energy via photosynthesis depends, among other factors, on the area of the cladodes remaining after cutting and the reserves for the next cycles [30, 31]. However, this response will be conditioned by the plant structure and the relationship between genotype, crop agroecosystem, and adopted management [31].

Regardless of harvest management and genotype, it is consistent to observe higher yields when primary or secondary cladodes are preserved (**Table 2**). This fact is related to a larger photosynthetic area that can provide faster growth and consequently higher productivity [30, 34]. In different states of the semiarid region of Brazil, it is common to observe harvest managements that preserve only the main cladode in search of a greater amount of cladodes per plant in the first harvest [31]. However, the plant will have fewer reserves for the next growth cycle, affecting later production.

Related to the ideal time for harvesting, [33] comment that the annual cut can be used as a management practice for forage cactus since the sum of fresh matter production and dry matter production can be greater when the annual harvest is adopted. However, it will also depend on other managements and cultural treatments adopted in addition to the selected genotype.

3. Cultural treatments (weeding, irrigation, fertilization)

3.1 Weeding and irrigation

The forage cactus planting in production units has been purposed for animal feed as forage in 98.5% [13]. When properly managed (improved varieties, density, organic fertilization, weed control, irrigation), forage cactus (*Opuntia* or *Nopalea*) will be able to produce enough forage to support 4–5 adult cows per ha/year [35].

Weed control, as an agronomic practice to reduce competition for nutrients, moisture, and light, is important to increase both green and dry biomass and crop water accumulation. Thus, it is possible to obtain a greater amount of forage, carrying capacity, and water reserve in the plants [21, 24]. The recommended control can be chemical or mechanical, but the most used control method in the Northeast of Brazil is cleaning with a hoe or mowing during the dry season. Chemically, the control is recommended from the early growth stage to minimize competition, although, in Brazil, there are no products registered for weed control for forage cactus [36]. There are few references regarding this topic (**Table 3**).

The use of irrigation for forage cactus is another of the agronomic practices considered. It is not a common practice, but in some regions where low precipitation associated with high night temperatures limits crop development, the application of small amounts of water can improve results in the planted area [21]. Thus, it is a technology that should be strategically used based on local rainfall, thermal regimes, and available clone [38]. The diversity of responses has been observed over time.

For species *Opuntia stricta* (Haw.) Haw., authors report irrigation depths of 355 mm to ensure fresh and dry matter production of forage cactus in the first production cycle [39]. However, irrigation depths between 1048 and 1090 mm can promote better crop responses in successive cycles [40]. Both cases in environments with an air temperature of 26.5 °C, and reference precipitation and evapotranspiration (ETc) of 354.7 and 2,072 mm, respectively. According to [41], *O. stricta* (Haw.) irrigated with up to 40% ETc (849 mm/year) and *Nopalea cochenillifera* (L.) Salm-Dyck (IPA-Sertânia and Miúda) with 80% ETc (1076 mm/year), can anticipate the harvest time of the crop concerning cultivation under rainfed conditions.

3.2 Fertilizing

The cacti grow in various types of soils and regions with rainfall between 300 and 600 mm annually, however, they are sensitive to high rainfall [42]. Saline soils are another limitation to the cultivation of the *Opuntia* and *Nopalea* because the growth of roots and shoots is reduced. [21] added that stress is caused when the concentration of sodium chloride (NaCl) reaches 25 mM reducing root development.

Clone	Control type	DMP (t/ha)	Reference
Gigante[1] Harvest 2 years	Chemical	11.9	[37]
	Manual labor (summer weeding and hoe)	4.93	
	No control	3.03	
Miúda[2] (0.5 × 0.5 m) Harvest 1 year	Manual labor	11.1	[24]
	No control	9.5	
Miúda[2] (1.0 × 1.0 m) Harvest 1 year	Manual labor	3.9	
	No control	4.5	

DMP: dry matter production
[1]*Opuntia ficus-indica* Mill.
[2]*Nopalea cochenillifera* (L.) Salm-Dyck.

Table 3.
Control of weeds used in forage cactus production.

Due to drought resistance and high efficiency in rainwater use, forage cactus planting is neglected in terms of soil fertility; which is a mistake. In semiarid regions and adequate climatic conditions, it is an unbeatable crop in terms of productivity and quality as an energy feed, for that it can be called The Queen of Forages in the Semiarid Region [43]. So, it must occupy the best fertile soil on the property.

As with all crops, the fertilization of forage cactus is conditioned to the fertility of the soil where it was or will be planted. Therefore, the first step to cultivate the forage cactus is the choice of the planting place, and the second to carry out the soil analysis. When the soil is submitted for analysis, the recommendation of fertilization for forage cactus is required. Or, with the analysis result, a professional can make the calculations to quantify enough limestone to correct soil acidity if necessary, and quantify the amount, formulate the planting and maintenance fertilizers for the crop.

In the nutritional aspect, it has long been recognized that forage cactus responds well to organic and chemical fertilization, as shown by [21, 42, 44, 45]. Also known the effect of the interaction between the level of fertilization, spacing, and environmental conditions of the crop influence the nutrients replacement. The higher population of plants more extraction of nutrients from the soil, and the greater requirement.

According to [42] forage cactus has a low nutritional requirement, but nutritional deficiency causes losses in yield and plant health. They report a quick response to the application of manure and chemical fertilizer in the production of new cladodes and fruits. Under greenhouse conditions, the application of 3–5 g/l of NPK (19:19:19) after fruit harvest was beneficial to the production of new cladodes. Another point reported by authors was the positive response to fertilization with tanned corral manure, which improves soil structure, nutrient availability, and soil water storage capacity. Thus, they recommend 6–10 t of barn manure/ha incorporated into the soil before planting.

In soil conditions, their recommendation is the application of 20 kg of N after harvesting cladodes, either for the production of *"nopalitos"* or forage, which indicates the need for constant nutrient replacement for the plant.

The recommendations above are for India and are contained in ICAR's Technical Bulletin No. 73, which still shows the recommendation by [46] with the combination of five tons of tanned corral manure and NPK (60:30:30)/ha at planting.

The five soil nutrients that may influence the *Opuntia* performance are N, P, K, B, and Na [47]. For [48] N, P, K, Ca, B, Mg, Fe, and Mn are the nutrients with the greatest effect on forage cactus growth in descending order, cited by [21].

Some research results for the states of Pernambuco and Paraíba prove the positive effect of fertilization with cattle manure on the *O. stricta* (Haw.) Haw and *N. cochenilifera* cv. "Miúda" yields (**Table 4**).

[51] suggested for South Africa the correction of the soil before the forage cactus planting intended for fruit production in dryland during summer rains. They indicated the ideal soil pH range of 6.5 to 7.5 and the fertilization indicated by soil analysis to obtain the soil nutrient levels as shown in **Table 5**.

Whereas the recommendation for forage cactus nutrition to produce fruits or "nopalito*s*" is inconsistent and contradictory, physiologically and morphologically different from many other crops [51], and discussed in other countries. The fertilization of forage cactus would be no different in Brazil. The indication of nutrients levels in the soil contained in **Table 5** can be used as an indicator to forage cactus fertilization in Brazil, where high dry matter productivity per area is expected. What is common where forage cactus is produced as an agricultural crop for fruit or forage is the use of fertilization to maintain productivity and perenniality.

Location	Plants/ha	manure (t/ha)	Increment (t/ha/2 years)	%	Reference
Parari, PB	20,000[1]	20	70.3 → 191.9(FM)	173	[49]
Bonito de Santa Fé, PB	20,000[2]	20	74.8 → 299.8(FM)	300	
Caruaru, PE	40,000[2]	30	9.6 → 42.6 (DM)	443.7	[50]

[1]*Orelha de Elefante Mexicana [Opuntia stricta (Haw.) Haw.].*
[2]*Nopalea cochenillifera (L.) Salm-Dyck cv Miúda; FM: fresh natter: DM: dry matter.*

Table 4.
Indicating that forage cactus responds positively to organic fertilization.

The great level of element in soil (mg/kg)			
P	K	Ca	Mg*
20–30	80–100	> 400	100–150

*Mg levels should not be bigger than Ca. Source: [52, 53] (adapted).

Table 5.
Suggested optimal soil nutrient levels for forage cactus fruit production in dryland summer crops in South Africa.

In Brazil, research about forage cactus retakes to the 1950s with agronomic trials on fertilization, planting spacing, and later on animal feed [21], and nowadays on irrigation, water salinity, and chemical weeding. Some studies indicate the composition and morphology of Brazilian Semiarid soils show diversity; they are vulnerable to degradation, due to the decrease in organic matter content and loss of fertility, and in arid, semiarid, and dry sub-humid climates it is characterized as desertification [28]. Data from INSA show that 9% of the Brazilian semiarid region is already desertified and 85% in a moderate process of desertification, a condition that makes the management of this soil more difficult and the need to use soil conservation and fertilization management techniques.

This diversity consists of shallow, stony, and sandy soils generally with low fertility in contrast to deeper soils with greater fertility. In some situations, saline soils are already found. [21] reported 19.2% of the soils in the Brazilian semiarid range from Litholic Neosols, shallow with an "A" horizon directly on the rock, to Latosols (21%), deep, well-drained, and with low organic matter content.

As we know the scope of forage cactus fertilization is generally neglected by producers. The reasons are many and generally, the areas chosen by the producers are characterized by their little agricultural vocation and usually with low fertility. [54] developed research with producers from Taperoá, PB, Brazil, and found that only 10% of producers performed soil analysis before planting forage cactus. However, 74% of the plantations were implanted in clayey soils, 20% in sandy-clay textured soils, and 6% cultivated cactus in sandy textured soils.

The search for greater productivity in the forage cactus crop has led researchers and producers to increasing plant density, increasing the number of plants per ha under cultivation. [55] indicate extraction of 0.9; 0.16; 2.58 and 2.35%, for N, P, K, and Ca, respectively by forage cactus cultivation indicating partial agreement with [42]. However, [56] cited by [21] demonstrated the positive effect on forage cactus production with increasing levels of organic fertilization and numbers of plants per ha in the state of Pernambuco. Even with a low level of nutrient requirement by forage cactus, the increase in dry matter production per area promotes high nutrient extraction per cultivated area causing the need for nutrient replacement after each harvest, whether annual (**Table 6**) or biannual. Logically, the amount of fertilizer needed to increase production will reach its limit.

Productivity (t DM/ha/year)	Nutrient annual removal (kg/ha)				
	N	P	K	Ca	Ratio t DM:Nutrient amount
5	45	8	129	117	1:1:1:1
10	90	16	258	235	2:2:2:2
20	180	32	516	470	4:4:4:4
40	360	64	1032	940	8:8:8:8
55	495	88	1419	1292	11:11:11:11
80	720	128	2064	1880	16:16:16:16

Calculated from [55]: extration of 0.9; 0.16; 2.58 and 2.35% for N, P, K e Ca from soil, respectively.

Table 6.
Nutrient extraction by forage cactus according to productivity.

Soil analysis	Implantation[1] (kg/ha)			Fertilizing[2] (kg/ha)		
Content in soil	Planting	Growth	After cutting	Planting	Growth	After cutting[2]
		Nitrogen (N)				
Do not consider		100	100		222	222
		Phosphorus (P_2O_5)				
P						
< 11 mg/dm^3	80	60	60	445		445
K		Potassium (K_2O)				
< 0.12 cmol$_c$/dm^3	100	60	100	167		167
Organic fertilization						
Cattle manure[3]	20,000					20.000

[1]*Fertilizing recommendation for the State of Pernambuco, Guide [58].*
[2]*Urea, Single superphosphate (P_2O_5) e potassium chloride (K_2O)*
[3]*Based on [50].*

Table 7.
Example of chemical and organic fertilization association for forage cactus based on hypothetical soil analysis and recommendation for the state of Pernambuco, Brazil.

Research by [57] showed the efficiency of organic fertilization decreased when using a low amount of cattle manure for planting with 160,000 plants/ha of forage cactus and recommended a minimum application of 40 t/ha every two years for this density. Greater productions occurred with the increase in population density and application of 80 t of cattle manure every two years, with values of 61; 90; 117 and 139 t DM/ha/two years, respectively, for planting densities of 20, 40, 80 and 160 thousand plants/ha.

Taking as an example a forage cactus planting in low fertility soil (P and K; **Table 7**), we used the fertilizer recommendation for forage cactus in Guide recommendation for crops in the state of Pernambuco.

4. The forage cactus as a invasive plant

[59] reported to have little information on the subject but asserts several occurrences of cactus becoming a problem as invasive plants in several countries around the world. According to him, species of commercial value such as *Opuntia*

ficus-indica and *Opuntia monacantha* have become invasive in several countries, requiring their control.

In Brazil, this is still not a problem be considered for cactus cladodes, however, [60, 61] cited by [62] comment cactus species are the most cosmopolitan and destructive among invasive plants in any parts of the world. Briefing, informative material from ICARDA – International Center for Agricultural Research in the Dry Areas reports after 150 years cultivation of *Opuntia ficus-indica* in South Africa reverted to its thorny form becoming an invasive plant and forming dense, impenetrable bushes with more than two million hectares invaded at the beginning of the 20th century, although, in the colder parts of the country, forage cactus was less aggressive and producers used it more extensively. Countries where the climate is more favorable such as Eritrea, Ethiopia, Yemen, Saudi Arabia, and Madagascar occurred a similar invasion.

The number of invasive species in South Africa has increased from 13, all *Opuntiae,* in 1947 to 35 in 2014, including at least eight *Cactoideae,* and some of them had to be subjected to chemical control followed by biological control if necessary [59].

5. Spinelles Cactus as forage and desertification mitigation

The semiarid in the world land structure is almost entirely characterized by a large number of small and medium sized family-owned establishments. In Brazil, 70% of the consumed food is produced by small producers [63]. Although family farming is economically in these regions crucial, producers in the semi-arid region are most vulnerable to the impacts of climate change. The combination of an adverse environment and economic activity that is dependent on nature leads to extreme vulnerability of the production systems, represented by virtual collapses under climatic conditions that are unfavorable to production. This, in part, results in economic fragility.

In dry areas around the world, periodic droughts have a major impact on rural properties, leading to serious socio-economic losses [29]. In these regions, biomass production is typically low (<5 tons of DM per ha per year), with low forage potential (<1 ton of DM per ha per year), leading to a low support capacity (12–15 ha to sustain an adult cow; Dubeux et al., 2015). However, producers should make efforts to identify and implement strategies to deal with these adversities, which can reward them with long-term resilience [64]. For this reason, [65] suggested corn crop for silage production. [66] evaluated five short cycle corn cultivars, recommended for silage production in semi-arid regions, and observed a productivity of 8.04 tons of DM/ha (6.12 to 9.68 tons of DM/ha).

However, the use of cactus, notably cactus cladodes (*Opuntia* and *Nopalea*), for ruminant feeding in dry areas has been increasing, as, for example, in North Africa [67] and northeast Brazil [68, 69]. Cactus is chosen for its high efficiency of water use, rapid dissemination, high water and energy content, and high biomass yield [70]. Recently, [71] suggested cactus *Opuntia stricta* (Haw.) Haw. cladodes as a new option for milk production in smallholder systems in semi-arid regions. In addition, [58] published productivity data of this cactus cladodes' clone in different semi-arid areas in Brazil and reported a minimum production of 40 tons of DM/ha and a maximum production of 60 tons of DM/ha, achieved every two years.

In general, energy is the most limiting "nutrient" for animal production. [72] showed that *O. ficus-indica* and *N. cochenillifera* has an average ME content of 2.34 Mcal/kg DM. In **Table 8** presents the estimates of DM productivity/ha of various forages that are commonly recommended for semiarid regions. Thus, they are equal

Item	Forages					
	Forage cactus	Sorghum silage	Alafafa	Leucaena	Buffel grass	Corn silage
ME (Mcal/Kg DM)	2.34	2.28	2.13	2.67	1.52	2.29
DM (ton/ha)	23.69	24.31	26.03	20.76	36.47	24.21

Table 8.
Metabolizable Energy (ME) content and productivity expectation of different forages.

to the potential for ME production/ha of forage cactus, which was 55,434 Mcal/ha (23,690 kg DM; 2.34 Mcal/kg DM). The average productivity of the forage cactus species was considered in the paper of [43]. The ME values of the various forages were taken from the Brazilian Tables of Feed Composition for Cattle [73].

It is impossible to achieve the productivity of the selected forages in semiarid conditions (**Table 8**) under low rainfall without irrigation. However, they should not be discarded, because they could be used, to a lesser extent in the diet, as a source of fiber.

Some other advantages justify spineless forage cactus use; for example, cows producing 15 kg of milk/day, fed with a diet contenting 50% of forage cactus, practically do not need water via a drinking fountain [74]. Spineless forage cactus is a perennial crop that allows for a reduction in implantation costs over time.

Due to its crude protein content (5.4%), CNF content (54.3%), and NDF content (24.8%), cactus cladodes combined with a cheap source of fiber (sugarcane bagasse, wheat straw) and NPN (urea), as a feeding strategy for ruminants, show very satisfactory results, including a reduction in the required amount of concentrated feed. [75] evaluated diets for crossbred lactating cows, with 61% forage cactus, 34.2% roughage, 1.7% urea, and only 3.1% soybean meal. They reported an average production of 11 kg milk/day. In another study, Holstein heifers, with an average weight of 243 kg, received a basal diet consisting of spineless forage cactus (69.8%), sugarcane bagasse (27.6%), and urea (2.6%), supplemented with 1 kg wheat bran per day. They showed an average gain of 0.71 kg/day [76]. Spineless forage cactus is an excellent feed for small ruminants. [77, 78] reported a positive performance for sheep with an average daily gain of 251 g/day, and lactating goats with average milk production of 2.97 L/day, respectively, when the animals were fed with spineless cactus.

A major issue that affects the global society is desertification, which is the process of land degradation in arid, semiarid, and sub-humid areas stemming from factors such as climatic variations and human activities [79]. Due to climatic conditions, soil characteristics, the inadequate exploitation of natural resources, and overgrazing, the Caatinga, a specific biome in Northeast Brazil, has become fragile and vulnerable [80]. In general, the causes of desertification in Northeast Brazil are not different from those typically found in other areas around the world. They are related to the exploitation of natural resources, to improper practices of land use (overgrazing and over-cultivation), and above all, to models of immediatism regional development [80].

It is necessary to consider the notorious contribution of livestock activity to the acceleration of the desertification process, along with the aforementioned climatic factor. According to [81], the use of semi-extensive or extensive livestock in semiarid areas becomes a factor in environmental changes due to the excessive stocking of animals in limits above the ecosystem's support. In the medium term, it exerts strong pressure on the floristic composition of the native vegetation due to the high palatability that is causing the extinction of species. It also exerts pressure on the soil due to the excessive trampling that causes compaction (in the rainy season)

Manipulation Level	DMY[**] (kg/year)	Available for animal intake	Forage cactus area (ha)
Nativa	4.000	400	0.02
Rebaixada	4.000	1600	0.08
Raleada	4.000	2400	0.13
Enriquecida	4.000	3600	0.18

[*] 20 tons of dry matter/year was considered.
[**] Dry matter yield.

Table 9.
Caatinga management and biomass production vs. forage cactus. [*]

and disintegration (in the dry season), which has negative effects on soil physical, chemical, and biological properties. In the long term, it contributes to the irreversible degradation of soils and vegetation, thus generating areas that are susceptible to the process of desertification.

The use of spineless forage cactus in areas where it can develop normally and may become the basis for ruminants' feed would increase the support capacity production systems. This would avoid the indiscriminate use of natural vegetation, mitigate desertification, and improve coexistence with the adverse conditions of the semiarid region. Taking Caatinga as an example that is an exclusive Brazilian biome with semiarid weather, vegetation with a few leaves and adapted to dry season, presents great biodiversities, but it is quite degraded by man.

According to [82], there are techniques for handling the Caatinga that can significantly increase the forage supply in that biome and contribute to its preservation. The main techniques used are thinning, lowering, and enrichment of the caatinga, with possible combinations between them. The thinning consists of making selective cuts in species of little forage and timber value, reducing the density of these plants in the area, thus allowing other species to develop and serve as a source of feed for the animals. Lowering is cutting the highest part of trees and shrubs to increase the forage supply for grazing animals. This practice makes forage in the pasture accessible, but it is not easily available because it has two meters high, becoming indicated for use in goat production systems or that combine goats and cattle. On the other hand, enrichment is a technique to improve forage production conditions by introducing perennial species. In addition to the benefits for herds, these management techniques help to regenerate native vegetation and optimize the use of forage resources (**Table 9**). There is a considerable increase in forage availability, from 400 (native caatinga) to 3600 kg of dry matter/ha/year (enriched caatinga).

Despite the increase verified with the manipulation of the Caatinga, it could be preserved using more productive species such as *Opuntia* and *Nopalea*, which would will produce much more in less area used fill less space. A comparison was made between the amount of dry matter in a hectare of native Caatinga or different management systems can make available to the animal and how much this would represent if forage cactus were used (**Table 9**). According to the simulation carried out, it can be seen that thousands of hectares of Caatinga could be preserved with the use of forage cactus. We must not forget that the forage cactus must be supplemented with fiber and nitrogen sources according to animal requirements.

6. Conclusion

Opuntia spp. and *Nopalea* spp. are cultivated and have been income sources for farmers as fruit, nutrition, medicine and forage use. Cultural treatments such as weeding control, irrigation and fertilization; stand and spacing are extremely important factors to consider in the planting of forage cactus in order to increase productivity.

Specifically for Brazil's semiarid region these species can make the difference as forage for animal feeding, cultivated as monoculture or intercropped, for soil conservation and desertification mitigation, source of water for animals, preservation of the Caatinga biome and be a potential source of income for producers if cultivated as vegetable for nutritional properties and medicinal derivative of fruits and cladodes for exports.

Author details

Marcelo de Andrade Ferreira[1*], Luciano Patto Novaes[2], Ana María Herrera Ángulo[3] and Michelle Christina Bernardo de Siqueira[1]

1 Federal Rural University of Pernambuco, Brazil

2 Federal University of Rio Grande do Norte, Brazil

3 Universidad Nacional Experimental del Táchira, San Cristóbal, Táchira, Venezuela

*Address all correspondence to: marcelo.aferreira@ufrpe.br

IntechOpen

References

[1] Souza, D.C.F., Lima, I.S., Santana, J.A., Almeida, A.Q., Gonzaga, M.I.S., Santana, J.F., 2018. Agroclimatic zoning palm forage (Opuntia sp.) for the state of Sergipe. Rev. Bras. Agric. Irrig. 12, 2338-2347. https://doi.org/10.7127/rbai.v12n100715

[2] Kaufmann, J.C., 2004. Prickly pear cactus and pastoralism in Southwest Madagascar. Ethnology. 43, 345-361. https://doi.org/10.2307/3774032

[3] Mayer, J.A., Cushman, J.C., 2019. Nutritional and mineral content of prickly pear cactus: A highly water-use efficient forage, fodder and food species. J. Agron. Crop Sci. 205, 625-634. https://doi.org/10.1111/jac.12353

[4] Morshedy, S.A., Mohsen, A.E.A., Basyony, M.M., Abdel-Daim, M.M., El-Gindy, Y.M., 2020. Effect of prickly pear cactus peel supplementation on milk production, nutrient digestibility and rumen fermentation of sheep and the maternal effects on growth and physiological performance of suckling offsprin. Animals. 10, 1-20. https://doi.org/10.3390/ani10091476

[5] Santos, D.C., Silva, M.C., Dubeux Júnior, J.C.B., Lira, M.A., Silva, R.M., 2013. Estratégias para uso de Cactáceas em zonas Semiáridas: Novas cultivares e uso sustentável das espécies nativas. Rev. Científica Produção Anim. 15, 111-121. https://doi.org/10.15528/2176-4158/rcpa.v15n2p111-121

[6] Silva, T.G.F., Araújo, J.T., Morais, J.E.F., Diniz, W.J.S., Souza, C.A.A., Silva, M.C., 2015. Crescimento e produtividade de clones de palma forrageira no semiárido e relações com variáveis meteorológicas. Rev. Caatinga. 28, 10-18.

[7] Silvera, K., Neubig, K.M., Whitten, W.M., Williams, N.H., Winter, K., Cushman, J.C., 2010. Evolution along the crassulacean acid metabolism continuum. Funct. Plant Biol. 37, 995-1010. https://doi.org/10.1071/FP10084

[8] Instituto Brasileiro de Geografia e Estatística (IBGE). Censo Agropecuário 2017. Rio de Janeiro: IBGE, 2017.

[9] Nefzaoui, A., 2018. *Opuntia ficus-indica* productivity and ecosystem services in arid areas. Italus Hortus. 25, 29-39. https://doi.org/10.26353/J.ITAHORT/2018.3.2939

[10] Vasconcelos, A.G.V., Lira, M.A., Cavalcanti, V.L.B., Santos, M.V.F., Willadino, L., 2009. Seleção de clones de palma forrageira resistentes à cochonilha-do-carmim (Dactylopius sp). Rev. Bras. Zootec. 38, 827-831. https://doi.org/10.1590/S1516-3598 2009000500007

[11] Borges, L.R., Santos, D.C., Cavalcanti, V.A.L.B., Gomes, E.W.F., Falcão, H.M., Silva, D.M.P., 2013. Selection of cactus pear clones regarding resistance to carmine cochineal *D. opuntiae* (Dactylopiidae). Acta Hortic. 995, 359-366. https://doi.org/10.17660/actahortic.2013.995.47

[12] Pinheiro, K.M., Silva, T.G.F., Sousa Carvalho, H.F., Santos, J.E.O., Morais, J.E.F., Zolnier, S., Santos, D.C., 2014. Correlations of the cladode area index with morphogenetic and yield traits of cactus forage. Pesqui. Agropecu. Bras. 49, 939-947. https://doi.org/10.1590/S0100-204X2014001200004

[13] Leite, M.L.M.V., Silva, D.S., Andrade, A.P., Pereira, W.E., Ramos, J.P.F., 2014. Characterization of forage cactus production in the Cariri region of Paraíba state - Brazil. Rev. Caatinga. 27, 192-200.

[14] Amorim, D.M., Silva, T.G.F., Pereira, P.C., Souza, L.S.B., Minuzzi,

R.B., 2017. Phenophases and cutting time of forage cactus under irrigation and cropping systems. Pesqui. Agropecu. Trop. 47, 62-71. https://doi.org/10.1590/1983-40632016v4742746

[15] Matos, L.V., Donato, S.L.R., Kondo, M.K., Lani, J.L., Aspiazú, I., 2021. Soil attributes and the quality and yield of 'Gigante' cactus pear in agroecosystems of the semiarid region of Bahia. J. Arid Environ. 185, 104325. https://doi.org/10.1016/j.jaridenv.2020.104325

[16] Amorim, P.L., Martuscello, J.A., Araújo Filho, J.T., Cunha, D.N.F.V., Jank, L., 2015. Morphological and productive characterization of forage cactus varieties. Rev. Caatinga. 28, 230-238. https://doi.org/10.1590/1983-212520 15v28n326rc

[17] Azevedo Junior, M.S., Ferreira Neto, M., Medeiros, J.F., Sá, F.V.S., Lima, Y.B., Lemos, M., Queiroz, J.L.F., Batista, R.O., 2020. Growth and biomass production of prickly pear in the second cycle irrigated with treated domestic sewage. Biosci. J. 36, 51-60. https://doi.org/10.14393/BJ-v36n1a2020-42175

[18] Ramos, J.P.F., Macêdo, A.J.S., Santos, E.M., Edvan, R.L., Sousa, W.H., Perazzo, A.F., Silva, A.S., Cartaxo, F.Q., 2021. Forage yield and morphological traits of cactus pear genotypes. Acta Sci. - Agron. 43, 1-11. https://doi.org/10.4025/ACTASCIAGRON.V43I1.51214

[19] Donato, P.E.R., Donato, S.L.R., Silva, J.A., Pires, A.J.V., Rosa, R.C.C., Aquino, A.A., 2016. Nutrition and yield of 'Gigante' cactus pear cultivated with different spacings and organic fertilizer. Rev. Bras. Eng. Agric. e Ambient. 20, 1083-1088. https://doi.org/10.1590/1807-1929/agriambi.v20n12p1083-1088

[20] Lopes, E.B., Brito, C.H., Albuquerque, I.C., Batista, J.L., 2010. Seleção de genótipos de palma forrageira (Opuntia spp.) e (Nopalea spp.) resistentes à Cochonilha-do-Carmim

(*D. opuntiae* cockerell, 1929) na Paraíba, Brasil. Eng. Ambient. 7, 204-215.

[21] Dubeux Jr., J.C.B., Ben Salem, H., Nefzaoui, A., 2017. Forage production and supply for animal nutrition, in: Crop Ecology, Cultivation and Uses of Cactus Pear. IX INTERNATIONAL CONGRESS ON CACTUS PEAR AND COCHINEAL CAM Crops for a Hotter and Drier World. FAO and ICARDA, Coquimbo, Chile, p. 73-92.

[22] Cavalcante, L.A.D., Santos, G.R. de A., Silva, L.M., Fagundes, J.L., Silva, M.A., 2014. Response of cactus pear genotypes to different crop densities. Pesqui. Agropecuária Trop. 44, 424-433. https://doi.org/10.1590/s1983-406320 14000400010

[23] Lédo, A.A., Donato, S.L.R., Aspiazú, I., Silva, J.A., Donato, P.E.R., Carvalho, A.J., 2019. Yield and water use efficiency of cactus pear under arrangements, spacings and fertilizations. Rev. Bras. Eng. Agric. e Ambient. 23, 413-418. https://doi.org/10.1590/1807-1929/agriambi.v23n6p413-418

[24] Ramos, J.P.F., Souza, J.T.A., Santos, E.M., Pimenta Filho, E.C., Ribeiro, O.L., 2017. Growth and Productivity of *Nopalea cochenillifera* in function of different planting densities in cultivation with and without weeding. Rev. Electron. Vet. 18, 1-12.

[25] Farias, I., Lira, M.A., Santos, D.C., Tavares Filho, J.J., Santos, M.V.F., Fernandes, A.P.M., Santos, V.F., 2000. Harvest managing and plant spacing of spinelles fodder cactus, under grain sorghum intercropping at the semi-arid region of Pernambuco state, Brazil. Pesqui. Agropecuária Bras. 35, 341-347. https://doi.org/10.1590/s0100-204x 2000000200013

[26] Peixoto, M. A., Carneiro, M.S.S., Amorim, D.S., Edvan, R.L., Pereira, E.S., Costa, M.R.G.F., 2018. Agronomic

characteristics and chemical composition of the forage palm for different cropping systems. Arch. Zootec. 67, 35-39. https://doi.org/10.21071/az.v67i257.3385

[27] Miranda, K.R., Dubeux Jr., J.C.B., Mello, A.C.L., Silva, M.C., Santos, M.V.F., Santos, D.C., 2019. Forage production and mineral composition of cactus intercropped with legumes and fertilized with different sources of manure. Cienc. Rural 49, 1-6. https://doi.org/10.1590/0103-8478CR20180324

[28] INSTITUTO NACIONAL DO SEMIÁRIDO–INSA. 2021 Campina Grande–PB, 2021.

[29] Marengo, J.A., Torres, R.R., Alves, L.M., 2016. Drought in Northeast Brazil—past, present, and future. Theor. Appl. Climatol. 129, 1189-1200. https://doi.org/10.1007/ s00704-016-1840-8.

[30] Lima, G.F.C., Rego, M.M.T., Dantas, F.D.G., Lôbo, R.N.B., Silva, J.G.M., Aguiar, E.M., 2016. Morphological characteristics and forage productivity of irrigated cactus pear under different cutting intensities. Rev. Caatinga. 29, 481-488. https://doi.org/10.1590/1983-21252016v29n226rc

[31] Matos, L.V., Donato, S.L.R., Da Silva, B.L., Kondo, M.K., Lani, J.L., 2020. Structural characteristics and yield of "Gigante" cactus pear in agroecosytems in the semi-arid region of Bahia, Brazil. Rev. Caatinga 33, 1111-1123. https://doi.org/10.1590/1983-21252020v33n426rc

[32] Lima, G.F.C., Silva, J.G.M., Dantas, F.D.G., Guedes, F.X., Rêgo, M.M.T., Aguiar, E.M., Lôbo, R.N.B., 2015. Effect of different cutting intensities on morphological characteristics and productivity of irrigated Nopalea forage cactus. Acta Hortic. 1067, 253-258. https://doi.org/10.17660/ActaHortic.2015.1067.35

[33] Ramos, J.P.F., Santos, E.M., Cruz, G.R.B., Pinho, R.M.A., Freitas, P.M.D., 2015. Effects of harvest management and manure levels on cactus pear productivity. Rev. Caatinga. 28, 135-142.

[34] Pereira, J.S., Cavalcante, A.B., Nogueira, G.H.M.S.M.F., Campos, F.S., Araújo, G.G.L., Simões, W.L., Voltolini, T., 2020. Morphological and yield responses of spineless cactus Orelha de Elefante Mexicana under different cutting intensities. Rev. Bras. Saúde e Prod. Anim. 21, 1-10. https://doi.org/10.1590/S1519-99402121142020

[35] Dubeux Jr., J.C.B., Santos, M.V.F., Mello, A.C.L., Cunha, M. V., Ferreira, M.A., Santos, D.C., Lira, M.A., Silva, M.C., 2015. Forage potential of cacti on drylands. Acta Hortic. 1067, 181-186. https://doi.org/10.17660/ActaHortic.2015.1067.24

[36] Santos, D.C., Farias, I., Lira, M., Santos, M.V.F., Arruda, G.P., Coelho, R.S.B., Dias, F.M., Melo, J.N., 2006. IPA. Documentos, 30. Manejo e Utilização da palma forrageira (Opuntia e Nopalea) p. 48.

[37] Farias, I., Melo, J.N., Dubeux Jr., J.C., Santos, M.V.F., Santos, D.C., Lira, M.A., 2001. Cactus forage productivy (*Opuntia ficus-indica* Mill) under diferent weed control methods, in: Reunião Anual Da Sociedade Brasileira de Zootecnia. SBZ, Piracicaba, p. 23-26.

[38] Araújo Júnior, G.N., Silva, T.G.F., Souza, L.S.B., Souza, M.S., Araújo, G.G.L., Moura, M.S.B., Santos, J.P.A.S., Jardim, A.M.R.F., Alves, C.P., Alves, H.K.M.N., 2021b. Productivity, bromatological composition and economic benefits of using irrigation in the forage cactus under regulated deficit irrigation in a semiarid environment. Bragantia. 80, 1-12. https://doi.org/10.1590/1678-4499.20200390

[39] Diniz, W.J.S., Silva, T.G.F., Jadna, M.S.F., Santos, D.C., Moura, M.S.B.,

Araújo, G.G.L., Zolnier, S., 2017. Forage cactus-sorghum intercropping at different irrigation water depths in the Brazilian Semiarid Region. Pesqui. Agropecu. Bras. 52, 724-733. https://doi.org/10.1590/S0100-204X201700 0900004

[40] Queiroz, M.G., Silva, T.G.F., Zolnier, S., Silva, S.M.S., Lima, L.R., Alves, J.O., 2015. Morphophysiological characteristic and yield of forage cactus under different irrigation depths. Rev. Bras. Eng. Agrícola e Ambient. 19, 931-938. https://doi.org/10.1590/1807-1929/agriambi.v19n10p931-938

[41] Araújo Júnior, G.N., Silva, T.G.F., Souza, L.S.B., Araújo, G.G.L., Moura, M.S.B., Alves, C.P., Salvador, K.R.S., Souza, C.A.A., Montenegro, A.A.A., Silva, M.J. da, 2021a. Phenophases, morphophysiological indices and cutting time in clones of the forage cacti under controlled water regimes in a semiarid environment. J. Arid Environ. 190, 1-9. https://doi.org/10.1016/j.jaridenv.2021.104510

[42] Kumar K., Singh D. and Singh R.S., 2018. Cactus pear: Cultivation and Uses. CIAH/Tech./Pub. No 73, p 38 ICA R-Central Institute for Arid Horticulture, Bikaner, Rajasthan, India.

[43] Rocha Filho, R.R., Santos, D.C.; Véras, A.S.C., Siqueira, M.C.B., Novaes, L.P., Mora-Luna, R., Monteiro, C.C.F., Ferreira, M.A., 2020. Can spineless forage cactus be the queen of forage crops in dryland areas?. J. Arid Environ. 186, p. 104426. https://doi.org/10.1016/j.jaridenv.2020.104426

[44] Lima, G.F.C., Wanderley, A.M., Guedes, F.X., Rego, M.M.T., Dantas, F.D.G., Silva, J.G.M., Novaes, L.P., Aguiar, E.M. 2015. Palma Forrageira irrigada e adensada: uma reserva Forrageira estratégica para o Semiárido Potiguar. EMPARN. Parnamirim, Rio Grande do Norte, Brasil.

[45] Zúñiga-Tarango, R., Orona-Castillo, I., Vázquez-Vázquez, C., Murillo-Amador, B., Salazar-Sosa, E., López-Martínez, J.D., García-Hernandéz, J.L., Rueda-Puente, E., 2009. Root growth, yield and mineral concentration of *Opuntia ficus-indica* (L.) Mill. under different fertilization treatments. J. Prof. Assoc. Cactus Dev. 11, 53-68.

[46] Kauthale, V., Aware, M. and Punde, K., 2017. Cactus an emerging fodder crop of arid and semiarid India. Booklet by BAIF Development Research Foundation, Pune.

[47] Nobel, P. S. 1995. Environmental biology. In: Barbera, G., Inglese, P., Pimientabarrios, E. Agro-ecology, cultivation and uses of cactus pear. Rome: FAO. p. 36-48.

[48] Baca Castillo, G.A., 1988. Deficiencias nutrimentales inducidas en nopal proveniente de cultivo in vitro. In: Reuniao nacional e international sobre conocimento y aprovechamento del nopal, 1 e 3, Saltilho. Memórias... Saltilho: Universidad Autonoma Agraria. p. 155-163.

[49] INSTITUTO NACIONAL DO SEMIÁRIDO–INSA. 2019. Palma forrageira: plantio e manejo. Campina Grande-PB, Brasil. p. 60.

[50] Souza, T. C. 2015. Sistemas de cultivo para a palma forrageira cv. Miúda (*Nopalea cochenillifera* Salm Dyck). 104 f. Thesis (Zootecnia). Universidade Federal Rural de Pernambuco, Recife, PE, Brasil.

[51] Potgieter, Johan and D'Aquino., 2017. Fruit production and post-harvest management. In: Crop Ecology, Cultivation and Uses of Cactus Pear. IX INTERNATIONAL CONGRESS ON CACTUS PEAR AND COCHINEAL CAM crops for a hotter and drier world. FAO FAO and ICARDA, Coquimbo, Chile, p. 51-71.

[52] Potgieter, J.P., 2001. Guidelines for the cultivation of cactus pears for fruit production. Fourth Revised Edition. Sinoville, South Africa, Group 7 Trust Printers. p 16.

[53] Potgieter, J.P., 2007. The influence of environmental factors on spineless cactus pear (Opuntia spp.) fruit yield in Limpopo Province. 120 f. MSc Thesis. University of the Free State Bloemfontein, South Africa.

[54] Souza, J. T.A., Farias, A. Aires, Lucena, J. N., Ferreira. T. C., O. S. J. C., 2013. Métodos de adubação orgânica e manejo do solo, na cultura da palma forrageira no cariri paraibano. Questões Contemporâneas. 12, 511-519. https://doi.org/10.12957/polemica.2013.8019

[55] Dubeux Júnior, J.C.B., Santos, M.V.F. 2005. Exigências nutricionais da palma forrageira. In: Menezes, R.S.C., Simões, D.A., Sampaio, E.V.S.B. (Eds). A palma no Nordeste do Brasil, conhecimento atual e novas perspectivas de uso. Ed. Universitária/UFPE. Recife, PE. p. 105-128.

[56] Souza, A. C., 1965. Novos experimentos com "palmas forrageiras" (*Opuntia ficus-indica*, Mill e *Nopalea cochenillifera*, (L.) Salm Dyck) em Pernambuco, Brasil. In: Congresso Internacional de Pastagens, 9, São Paulo, Anais. São Paulo: Secretaria da Agricultura do Estado de Pernambuco, 2, 1465-1469.

[57] Silva, N.G.M. 2012. Produtividade, morfometria e acúmulo de nutrientes da palma forrageira sob doses de adubação orgânica e densidades de plantio. 97 f. Thesis (Zootecnia). Universidade Federal Rural de Pernambuco, Recife, PE. Brasil.

[58] Santos, D. C., Lira, M. A., Dubeux júnior, J. C. B., Santos, M. V. F., Mello, A. C. L. Palma forrageira. In: Cavalcanti, F. J. A. (Coord.). Recomendações de adubação para o estado de Pernambuco:

2ª aproximação. 3.ed. Recife: Instituto Agronômico de Pernambuco–IPA, 2008. p. 178.

[59] Zimmermann, Helmuth., 2017. Global invasions of cacti (Opuntia sp.): control, management and conflitcs of interst. In: Crop Ecology, Cultivation and Uses of Cactus Pear. IX INTERNATIONAL CONGRESS ON CACTUS PEAR AND COCHINEAL CAM crops for a hotter and drier world. FAO FAO and ICARDA, Coquimbo, Chile, p. 171-185.

[60] Cronk, Q.C.B. & Fuller, J.C.,1995. Plant invasions: the threat to natural ecosystems. Chapman & Hall, London.

[61] Bright, C., 1998. Life out of bounds. The Worldwatch Environmental Series. Norton, New York.

[62] Zimmermann, H. G.; Moran, V. C. M and Hoffmann. 2001. The renowned cactus moth, cactoblastis cactorum (lepidoptera: pyralydae): its natural history and threat to nativa opuntia floras in mexico and united states of america. Fla. Entomol. 84, 543-551.

[63] MDA, 2017. Ministério da Agricultura Familiar e do Desenvolvimento Agrário (MDA), Brasil: 70% dos alimentos que vão à mesa dos brasileiros são da agricultura familiar. Brasília, Brasil. http://www.mda.gov.br/sitemda/noticias/brasil-70-dos-alimentosque-v%C3%A3o-%C3%A0-mesa-dos-brasileiros-s%C3%A3o-da-agricultura-familiar, Accessed date: 20 March 2021.

[64] Berman, R., Quinn, C., Paavola, J., 2012. The role of institutions in the transformation of coping capacity to sustainable adaptive capacity. Environ. Dev. 2, 86-100. http://dx.doi.org/10.1016/j.envdev.2012.03.017

[65] Vitória, E.L., Longui, F.C., Fernandes, H.C., Guimarães Filho, C.C., 2011. Influência do tipo de preparo do

solo e velocidade de semeadura em características agronômicas da cultura do milho. R. Agrotec. 2, 44-52. http://dx.doi.org/10.12971/2179-5959.v02n02a04

[66] Neves, A.L.A., Santos, R.D., Pereira, L.G.R., Tabosa, J.N., Albuquerque, I.R.R., Neves, A.L.A., Oliveira, G.F., Verneque, R.S., 2015. Agronomic characteristics of corn cultivars for silage production. Semina: Ciênc. Agrár. 36, 1799-1806. https://doi.org/ 10.5433/1679-0359.2015v36n3 Supl1p1799

[67] Mahouachi, M., Atti, N., Hajji, H., 2012. Use of spineless cactus (Opuntia ficus indica) for dairy goats and growing kids: impacts on milk production, kid's growth, and meat quality. Sci. World J. https://doi.org/10.1100/2012/321567.321567 ID.

[68] Lins, S.E.B., Pessoa, R.A.S., Ferreira, M.A., Campos, J.M.S., Silva, J.A.B.A., Silva, J.L., Santos, S.A., Melo, T.T.B., 2016. Spineless cactus as a replacement for wheatbran in sugar cane-based diets for sheep: intake, digestibility, and ruminal parameters. R. Bras. Zootec. 45, 26-31. https://doi.org/10.1590/S1806-9290201600010 0004

[69] Monteiro, C.C.F., Melo, A.A.S., Ferreira, M.A., Campos, J.M.S., Souza, J.R.S., Silva, E.T.S., Andrade, R.P.X., Silva, E.C., 2014. Replacement of wheat bran with spineless cactus (Opuntia fícus indica Mill cv Gigante) and urea in the diets of Holstein x Gyr heifers. Trop. Anim. Health Prod. 46, 1149-1154. https://doi.org/10.1007/s11250-014-0619-0

[70] Ben Salem, H., 2010. Nutritional management to improve sheep and goat performances in semiarid regions. R. Bras. Zootec. 39, 337-347. https://doi.org/10.1590/S1516-35982010001300037

[71] Silva, R.C., Ferreira, M.A., Oliveira, J.C.V., Santos, D.C., Gama, M.A.S.,

Chagas, J.C.C., Inácio, J.G., Silva, E.T.S., Pereira, L.G.R., 2018. Orelha de Elefante Mexicana (*Opuntia stricta* [Haw.] Haw.) spineless cactus as an option in crossbred dairy cattle diet. S. Afri. J. Anim. Sci. 48, 516-525. http://dx.doi.org/10.4314/sajas.v48i3.12

[72] Rocha Filho, R.R., Santos, D.C., Véras, A. S. C., Siqueira, M.C.B., Monteiro, C. C. F., Luna, R.M., FARIAS, L. R., Santos, V.L.F., Chagas, J. C. C., Ferreira, M.A., 2021. Miúda (*Nopalea cochenillifera* (L.) Salm-Dyck) is the best genotype of forage cactus for feeding lactating dairy cows in semiarid regions. Animals. 11, p. 1-13. https://doi.org/10.3390/ani11061774

[73] Valadares Filho, S.C., Lopes, S.A., Machado, P.A.S., Chizzotti, M.L., Amaral, H.F., Magalhães, K.A., Rocha Junior, V.R., Capelle, E.R., 2021. Tabelas Brasileiras de Composição de Alimentos para Bovinos. CQBAL 4.0. [www Document].

[74] Lima, R.M.B., Ferreira, M.D.A., Brasil, L.H.A., Araújo, P.R.B., Véras, A.S.C., Santos, D.C. dos, Cruz, M.A.O.M., Melo, A.A.S., Oliveira, T.N., Souza, I.S., 2003. Substituição do milho por palma forrageira: comportamento ingestivo de vacas mestiças em lactação. Acta Sci. Anim. Sci. 25, 347-353. https://doi.org/10.4025/actascianimsci.v25i2.2029

[75] Wanderley, W.L., Ferreira, M.A., Batista, A.M.V., Véras, A.S.C., Santos, D.C., Urbano, S. A., Bispo, S.V., 2012. Silagens e fenos em associação à palma forrageira para vacas em lactação. ˜ Consumo, digestibilidade e desempenho. Rev. Bras. Saúde e Prod. Anim. 13, 745-754. https://doi.org/10.1590/s1519-99402012000300014

[76] Carvalho, M.C., Ferreira, M.D.A., Cavalcanti, C.V. de A., Lima, L.E. de, Silva, F.M. da, Miranda, K.F., Chaves Véras, A.S., Azevedo, M. De, Vieira, V.D.C.F., 2005. Associação do bagaço de

cana-de-açúcar, palma forrageira e ureia com diferentes suplementos em dietas para novilhas da raça holandesa. Acta Sci. Anim. Sci. 27, 247-252. https://doi. org/10.4025/actascianimsci.v27i2.1229

[77] Lopes, L.A., Carvalho, F.F.R., Cabral, A.M.D., Batista, M.V., Camargo, K.S., Silva, J.R.C., Ferreira, J.C.S., Pereira Neto, J.D., Silva, J.L., 2017. Replacement of tifton hay with alfalfa hay in diets containing spineless cactus (*Nopalea cochenillifera* Salm-Dyck) for dairy goats. Small Rumin. Res. 156, 7-11. https://doi.org/10.1016/j.smallrumres. 2017.08.006

[78] Moura, M. de S.C., Guim, A., Batista, A.M.V., Maciel, M. do V., Cardoso, D.B., Lima Júnior, D.M. de, Carvalho, F.F.R. de, 2020. The inclusion of spineless cactus in the diet of lambs increases fattening of the carcass. Meat Sci. 160, 107975. https://doi.org/10. 1016/j.meatsci.2019.107975

[79] Brasil, 1997. Resolução CONAMA N° 238, de 22 de dezembro de 1997. Brasil.

[80] Araújo, C.S.F., Sousa, A.N., 2011. Estudo do processo de desertificação na Caatinga: uma proposta de educação ambiental. Ciência Educ. 17, 975-986. https://doi.org/10.1590/s1516-731320 11000400013

[81] Parente, H.N., Parente, M.O.M., 2010. Impacto do pastejo no ecossistema caatinga. Arq. Ciência Veterinária Zool. 13, 115-120.

[82] Silva, N.L., Araújo Filho, J.A., Sousa, F.B., 2007. Manipulação da vegetação da caatinga para produção sustentável de forragem. Sobral: Embrapa Caprinos, 2007: Circular técnica, 34. p. 11.

Chapter 9

Sewan Grass: A Potential Forage Grass in Arid Environments

Sanjay Kumar Sanadya, Surendra Singh Shekhawat and Smrutishree Sahoo

Abstract

Sewan grass (*Lasiurus scindicus*), a popular pastoral species, is getting some much-needed attention as mechanization, modernity in agriculture, and illicit grazing pose severe risks to biodiversity conservation in arid and semi-arid areas. It is found mainly in wastelands, dunes, hammocks, and sandy plains but less popular for cultivation in farmer's fields. Sewan grass has many features like good nutritional value, soil binder, tolerance to high temperature, high digestibility and palatability, and prolonged drought conditions contributed greatly towards its success as a potential forage species in arid environments. It contains significant quantities of crude fibres, lignin, minerals and crude protein, and varies in the proportion of their tissue that can be digested by ruminants. Most research focuses on the species as a forage plant and agronomical practices and is largely published in agricultural and grey literature. Meanwhile, there is a lack of information about breeding strategies and seed production technologies. Therefore, here we present a comprehensive review about agronomic management, breeding, and seed production strategies systematically that will aid in the management of sewan grass now and into the future.

Keywords: Arid environments, Breeding methodologies, Diversification, Nutritional quality, Thar Desert

1. Introduction

Indian hot arid zone covers an area of 32 million ha called 'Thar Desert'. 85 percent of the hot Desert lies in India and the rest of the 15 percent in Pakistan. It represents the most inhospitable arid zone of the world spreading mostly in the states of Rajasthan, Gujarat, Punjab, Haryana, Karnataka, and Andhra Pradesh in India. About 91 percent of the Indian desert falls in Rajasthan covering about 61 percent geographical area of the state. The Aravali hills intersect Rajasthan to the Northeast (semi-arid) and in the West lies the Great Indian Desert 'Thar'. High wind velocity, huge dune, semi-stabilized and stabilized dunes of different nature, high diurnal variation in temperature, scanty and poor rainfall, intense solar radiation, and high rate of evaporation are the main characteristics of the Thar Desert. The natural grasslands lie in Desert areas are highly deteriorated stage with the productivity of only 300–400 kg/ha/year. *Dichanthium-Cenchrus-Lasiurus* type grasslands are associated with sub-tropical, arid, and semi-arid regions comprising the northern portion of Gujarat and the whole of Rajasthan excluding the Aravalli ranges in the South, western Uttar Pradesh, Punjab, Haryana, and Delhi State between 23 and 32°N and 68 and 80°E. The principal perennial grass species of such grasslands are buffel-grass (*Cenchrus ciliaris*), birdwood grass (*Cenchrus setigerus*), marvel grass (*Dichanthium annulatum*), khavi grass (*Cymbopogon jawarancusa*), bermuda grass (*Cynodon dactylon*), wire grass (*Eleusine compressa*), sewan grass (*Lasiurus scindicus*), pan dropseed (*Sporobolus marginatus*), tantia (*Dactyloctenium sindicum*), halfa grass (*Desmostachya bipinnata*) etc. [1]. The dominant perennial grass *i.e.* indigenous sewan grass is popularly known as the "King of Desert grasses". Sewan grass (*Lasiurus scindicus* Henr.) belongs to the family *Poaceae* is native to dry areas of North Africa, Sudanese and Sahelian regions, East Africa and Asia. It is highly tolerant to drought but should be protected from the wind in the early stages of the establishment [2].

Sewan grass is a bushy, hairy inflorescence, multi-branched, C_4 desert grass and a stout woody rhizome [2, 3] find in wastelands of arid region. The wild form of Sewan grass (*Lasiurus hirsutus*) is a diploid species having somatic chromosome number (2x) 20 however some species of grass are vary with chromosome numbers and polyploidy nature also [4]. Sewan grass is a perennial grass that can live up to 20 years. Fertilization is not necessary because it can be grown through vegetative propagating material such as root slips. Sewan grass forms bushy thickets in sandy deserts where it is used for pasture, hay, and fodder for livestock. It is found in dry open plains, rocky ground, and gravelly soils [5]. It is relished by ruminants but does not stand heavy grazing and disappears when overgrazed [6].

Comparative performance of major grasses (sewan grass, marvel grass, buffel grass, birdwood grass and bermuda grass) of arid region are presented in **Table 1**. Sewan grass has a higher calcium content and lignin than other grasses, such as marvel grass, buffel grass, birdwood grass and bermuda grass. The components of crude fiber are cellulose, lignin and hemicellulose. However, in case of other nutritional properties sewan grass has lower than other grasses but due to its drought resistance ability can grown in very low rainfall condition (lower than 250 mm) and useful for small ruminants such as sheep and goat.

Nutritional quality	Sewan grass	Marvel grass	Buffel grass	Birdwood grass	Bermuda grass
Dry matter (%)	30–33	31–33	28–30	30–32	29–31
Crude protein (%)	6–7	5–6	6.5–7.0	6–7	9–10
Crude fibre (%)	35–55	35–45	38–42	39–40	29–31
NDF (%)	75–77	76.1	75.1	74.0	66.7
ADF (%)	45–49	47.6	46.6	45.0	36.7
Hemicellulose (%)	30–32	28.5	28.5	29.0	30.0
Lignin (%)	7.3	7.1	6.0	6.6	4.7
Ash content (%)	8–9	9.6	9.1	11.0	9.5
Potassium (g/kg DM)	9.5	11.2	19.5	19.0	15.0
Calcium (g/kg DM)	5.1	3.4	2.6	3.8	4.5
Magnesium (g/kg DM)	1.0–2.5	1.1	2.2	2.5	1.8
Phosphorous (g/kg DM)	0.2–1.6	1.6	1.7	1.9	2.2
Organic matter digestibility	52–55	55.1	56.7	57.0	58.4
Energy digestibility	50–52	52.7	54.2	54.9	55.8

Table 1.
Nutritional quality of sewan grass with other major grasses of arid environment.

Thirty days cutting interval at a height of 15 cm gives the best fresh fodder and dry matter yields. Sewan grass yields 2.7 to 10.5 tonnes fresh forage/ha/year and up to 3.4 tonnes DM/ha in well-established swards [3]. The low yield can be improved by annual seeding of companion legumes such as guar bean (*Cyamopsis tetragonoloba*) or moth bean (*Vigna aconitifolia*) [7, 8]. Sewan grass is very important in arid environments because it covers soil especially at the top 15 to 30 cm that helps to protect soil transportation or soil erosion [9], and improve soil health due to the continued decaying of roots of the grass. It can be used to stabilize desert dunes and hummocks [2, 3]. In deteriorated rangelands of Saudi Arabia, sewan grass helps to control the low value invasive species *Rhazya stricta* by smothering its seedlings. It is a useful tool to improve rangeland management [10]. However, sewan grass tolerates prolonged droughts, but has not been found growing in higher rainfall zones and faces a serious threat of becoming an endangered species due to changes in the land use pattern and overgrazing [11]. Reseeding arid rangelands with species such as *Lasiurus scindicus* were found more palatable than its native species *Lasiurus hirsutus* and improved the forage resources at degraded Dera Ghazi Khan Rangeland in Pakistan [12]. Sewan grass is a palatable grass for goat, sheep, and camel, but supplementation is required to meet their nutritional requirements [13–15]. Supplementation with crushed guar seeds (*Cyamopsis tetragonoloba*) at 150 g/head increased DM intake and diet digestibility in ewes grazing sewan grass [16]. The studies with different vegetations growing on the wastelands and grazing lands showed that the association of sewan with other vegetations depends on the area and rainfall pattern of the zone. In the Jaisalmer district, its association has been seen with *Elusine compressa* whereas, in Bikaner, it also comes well with *Cymbopogon jwarancusa*. Over the years, people of the desert have evolved a lifestyle around the sewan grass, based on animal care.

2. Distribution

In the world, sewan grasslands are mostly found in dryland areas such as African countries, arid and semi-arid regions of Asia, South America and Europe. Sewan grass is mainly grazed by ruminants, generally in association with *Cenchrus ciliaris* and *Cenchrus setigerus*, which occupy the same agro-ecological niche, especially in Rajasthan and Pakistan [12, 17, 18]. In India, sewan grass covers approximately 0.1 million hectares of the area including western Rajasthan, Uttar Pradesh, Haryana, Punjab, some parts of Delhi, and Gujarat [1]. The sewan is the most suitable and occurring species in 18–28 sub-zones of the western Rajasthan. In the western Rajasthan state of India, The main distribution zone starts from the west of Jodhpur to Barmer districts towards Bikaner. The hummocky sandy, plains of Bikaner and Barmer or adjoining districts also support the extensive sewan grasslands. Until the last decade, about 80% of the total geographical area of Jaisalmer district covering Nachana, West Puggal, Mohangarh, Sultana, and Binjewala supported sewan grasslands. The Sriganganagar, and Hanumangarh districts are suitable for Agri-silvi-pasture system with special preference to the sewan as a component.

3. Climatic conditions in Sewan Grasslands

3.1 Rainfall

The high inter annual variation of rainfall is the single major factor influencing the agricultural production in the region. The mean annual rainfall in western

Rajasthan received from 100 to 400 mm in the arid region of Rajasthan with a coefficient of variation of 40–70 per cent. More than 90 per cent of total rainfall is received in rainy season. In these parts, perennial grasses play major role in the economy of rural masses as well as survival of large cattle population. The areas receiving annual rainfall from 100 to 300 mm/year are the main locations where the natural sewan grass exists. The rainfed cropping zone is the main growing zone (More than 80 per cent) of sewan on the interdunal plains.

3.2 Temperature

The desert stands for extremes of temperature ranging from −5.7°C during winters to 48°C during summers. During winters mean maximum temperature varies from 24–26°C parts with the highest mean temperature of 33.3°C in western part of the region. January is the coldest month mean minimum temperature of 6.5 to 9.5°C. During summers, the mean maximum temperature varies from 36.1°C in east to 38°C in the west.

3.3 Drought

The frequency and occurrence of droughts in arid region are much higher than other regions in drought affected states in India. Out of 13 states repeatedly declared as drought-prone, Rajasthan is the most critical state in the country with highest probabilities of drought occurrence and rainfall deficiencies. Several records shows that about 48 drought years have been reported of varied intensity since 1901 in last 102 years and only 9 years out of them were totally free from drought [19]. The impact of recurrent droughts is the less hazardous than the consecutive droughts of 3–4 years (1984–1987). Consecutive droughts affect the sewan fresh fodder production very badly leading to mortality of animals. The studies conducted at CAZRI, Jodhpur revealed that the sewan could survive under extreme arid and severe drought conditions below 250 mm annual rainfall [20]. It has also been observed that the probability of experiencing severe droughts affecting the grass production in a rainfall zone below 200 mm is about 50 per cent. Sewan being a promising desert grass provides sustained forage production for a longer period even under the harsh climate or lean period of arid regions of western Rajasthan.

3.4 Landforms

Sewan has been found to be more suitable for wind strip cropping as an associate component of silvi-pasture system in the areas of sand dunes and undulations of sandy undulating aggraded alluvial, interdunal plains and sandy undulating buried piedmonts. Fourteen major landforms have been identified in Rajasthan as a whole. Among them, mainly the deposited sandy undulating plains are found to be more appropriate for sewan grass coverage and growth.

3.5 Existing situation

A large area as wastelands is available in arid region, which can be utilized for development of grasslands and establishment of pastures. The estimates showed that in India available wastelands vary from 56.60 million ha (17.21 per cent of geographical area of the country i.e. 328.72 Mha). The technologies for improvement and management of pastures, grass-legume mixed pastures and silvipastoral system to increase the carrying capacity of grazing lands is available [21].

3.6 Production technology

The production technologies of sewan grass are different from other arable crops grown under rainfed condition of western Rajasthan because of perennial nature (up to 15 to 20 years after sowing). The grass takes 2–3 years of its development to attain optimum yield and continued for 15 years or more. Generally, the non-cultivable wastelands used for production of sewan grass, which fall under class VIII of Land Use Capability Classification. Packages and practices to be use for production of sewan grass are described in **Table 5**. To mitigate the effect of drought and moisture scarcity adoption of soil-moisture conservation measures are very crucial tools that are presented in **Table 2**.

3.7 Seed production technology

The quality seed production of grasses is a challenging task for the breeders and agronomist. The major constraint in development of pasture is supply of inadequate, poor quality seed. In our country, the requirement of tropical range grasses and legumes is about 3000 t/year whereas the supply is only about 450 t/year having a very large gap between demand and supply. Many problems are associated with the grass seed production. Very low effort has been made to develop the high yielding grass varieties. It has also been noticed that the high fodder yielding varieties are very poor seed yielder. The maintenance of seed purity is also difficult due to its perennial nature and tussock making habit. The seed maturity in sewan grass is unsynchronized. The seed production is vulnerable to adverse weather conditions i.e. windstorms, rainfall, drought etc. It has been noticed that the high wind velocity leads to the mature seeds to fall on the ground and occasional heavy rainfall destroy the seeds. The occurrence of drought or moisture scarcity results in lower seed production. Under these circumstances, seed production opportunity and its exploration are very poor, which restricts popularization of sewan seed production among the farmers. Due to the unsynchronized maturity, it takes seven to ten days for all the spikelet to mature in normal season and it may extend up to 20 days in cooler months [31]. As per Indian Minimum Seed Certification Standards

Measures	Remarks	References
Construction of contour furrows	60 cm wide x 25 cm deep and distance of 10–15 m across the slope. Increases the fodder production up to 130 per cent.	[30]
Inter row water harvesting (IRWH) system	30 cm wide raised ditches are alternated with 70 cm of wide raised bed improves the soil moisture status in the field. Seeds sown on the edge of the ditches increased forage to 66 per cent over conventional system of planting.	[30]
Intercultural operations after 20–30 days of sowing	The most effective and common practice in the field. This practice removes weeds, reduces the loss of water through weeds and the fine particles dispersed on the soil surface by intercultural operation work as a surface dust mulching check the water loss from the soil. The intercultural operation breaks the capillaries and stop water evaporation from the soil which ultimately becomes available to the plants for longer period to the grass.	[30]

Table 2.
Soil-Moisture conservation measures.

Field standards	IMSCS	
	Foundation seed (FS)	Certified seed (CS)
Field standards		
Isolation distance (m)	20	10
Field inspection (nos.)	3	3
Off-type plants (%)	0.10	1.0
Inseparable other crop plants (nos.)	None	None
Objectionable weed plants (nos.)	None	None
Designated diseases (nos.)	None	None
Designated pests (nos.)	None	None
Seed standards		
Minimum Physical purity (%)	80.0	80.0
Minimum Genetic purity (%)	99.0	98.0
Maximum Inert matter (%)	20.0	20.0
Maximum other crop seed (nos./kg)	20	40
Maximum Other varieties seed (nos./kg)	20	10
Fields of the same variety not conforming to varietal purity requirements for certification (nos./kg)	20	20
Fields of another *Lasiurus* spp. known to cross or suspected of being able to cross (nos./kg)	200	200
Maximum Total weed seed (nos./kg)	20	40
Maximum Objectionable weed plants (nos./kg)	None	None
Submitted sample size (gm)	200.0	200.0
Working sample size (gm)	20.0	20.0
Maximum Moisture per cent	12.0	12.0
Minimum Germination per cent	20.0	20.0
For vapour-proof containers per cent	8.0	8.0

Table 3.
*Field and Seed standards for Sewan grass (*Lasiurus scindicus*) as per IMSCS.*

(IMSCS), Field and seed standards for identification, release sewan grass cultivars are mentioned in **Table 3**.

3.8 Seed collection

Different methods have been applied to collect sewan grass seed with less effort:

a. Cutting or collection when upper 25 per cent spike has been matured whole spike with its stem or when 75 per cent of the spike are matured and dried up to 2–3 days then the seeds were collected from spikes. The seed harvest of these methods has 35–40% germination, which satisfactory.

b. Another method introduced by CAZRI, Jodhpur in which caryopses of the spikes harvested by following hand cutting spike heads at optimum time termed as modified method. Seeds were also collected manually as per maturity called traditional method. The results revealed that mean seed germination percentage of October and November harvests was at par in both the methods while in March harvest this was almost double to the traditional method. However, germination was more in the seed harvest of traditional method. Hence modified method is less labor intensive and cost effective [22].

c. In the forest areas, seed is collected manually from the ground, these seeds have very poor germination percentage due to damage caused by ants. Hand collection as per maturity seed provides quality seed of good germination.

4. Agronomical principles for sewan grass seed production

The site should have all the agro eco-characteristics, which can help in growth, development and management of grass stand. If drought occurs at the time of seed formation, there should be provision of life saving irrigation for quality seed production. The sowing methods, fertilizer application etc. have to be followed as per the practices recommended for grass production (**Table 4**). Under irrigated condition good quality seed could be produced except under low temperature conditions in December and January. Grass should be harvested in the active rainy season to avoid the losses to seed due to rains. If cuts have been taken in July–August, from September onward there will be profuse tillering and more inflorescence production.

5. Sewan in alternate land use system

Alternate land use system is appropriate in areas where subsistence farming is practiced in fragile ecosystems and it poses more potentiality and flexibility in land use than the traditional crop production systems. An ideal system for dry land areas should have a judicious mix of crops, trees and grasses only then the natural resources will be judiciously utilized and returns maximized without any detrimental effect to environment [43]. Different alternate land use systems have classified in arid environments *viz.* Horti-Pasture system, Silvi-Pasture system, Agro-Forestry system, Agri-Pasture system, Agri-Horticultural system, Horti-Silvipasture and Agri-silviculture. Out of them, Agri-Pastoral system, Horti-Pastoral system and Silvi-Pastoral system are found very effective systems in which sewan grass use as alternate crop or grass to give maximum benefits (**Table 5**).

6. Land diversification and value addition

As we have already discussed that sewan grass lives more than 20–25 yielded up to 10–15 years but due to modernization in agriculture i.e. heavy grazing, mechanization, and economically important crop dependency of farmers sewan grass is being disappeared from the farmer's field and limited at wasteland areas. Therefore, there is need to conserved sewan grass and continuously supply sewan forage to the livestock. That can possible through land diversification that means to use land efficiently by growing sewan grass with arable crops without affecting the yield of both grass and arable crops. Strip cropping of sewan grass with arid legumes helps to conserved and maintains yields of both crop and grass. Another way to utilize the non-cultivable forest areas the planting sewan by adopting advanced production technologies and soil moisture measures. Thus, sewan can be used in diversification

Package & Practices	Description
Environmental features	
Soil	Sewan grass performs well on alluvial sandy plains, low dunes, hummocks and light textured soils with pH 8.5.
	In this type of soil, upper horizon is calcareous but quantity of $CaCO_3$ increases with down profile.
Climate	The climate of sewan-dominated zone has low and erratic rainfall (below 250 mm) and high temperatures.
	The aridity index is 250 whereas the Thornthwaite moisture Index value is below −40 for sewan grass growing areas.
	During summer season, temperature should be up to 45°C and in winter season below −3°C.
Agronomic practices	
Land preparation	At the initial stage of growth of grass requires ploughing is essential to make field free from weeds.
	The land should be properly ploughed once by disc followed by harrow to avoid the termite infestation and favors better establishment of sewan seedlings.
	To protect the pastures from illicit grazing should be fenced properly that called in local language *Jharberi* or *bordi*.
Varieties	Mostly landraces are dominated in the pasturelands and forest areas of western Rajasthan, Central Arid Zone Research Institute (CAZRI), Jodhpur and its research centers have taken a lead to developed sewan grass varieties. Varieties *viz.* CAZRI Sewan-1 (CAZRI 30–5), Jaisalmeri Sewan (RLSB 11–50), CAZRI 317 and CAZRI 319 have been released in last two decades (2000–2020).
Seed Treatment	To obtain better germination and save the seed from the attack of pests, seeds should be in fresh water for 3 hours and then wash with tap water for about 15 minutes [32].
	The insecticide such as BHC or Aldrin powder can be mixed with the mixture to protect the seed from insects after sowing [33].
	Seed germination or seed setting can increase foliar spray of combination of Cycocel (100 ppm) and Pactobutrazol (200 ppm) [34].
	It has been confirmed that treatment with 0.2% KNO_3 gave significantly higher germination (20.9%) than control (18.0%).
Sowing	Test weight of sewan grass is 7 g, which make as the seeds vulnerable to winds. Therefore, care should be taken for better placement of seed. Generally, two methods are recommended for sowing:-
a) Furrow sowing	In this method, seeds mix with moist sandy soil in 1:5 ratios in such a way that in one crunch of mixer approximately 10–12 seeds should be available for sowing.
	Furrows opened with the help of tractor or desi plough and drilled the mixtures in 2-3 cm depth with 75 to 100 cm spacing then covered with soil layer to avoid the instant loss of soil moisture from furrows and safety the seed from ants and other biotic agents [35, 36].
	The intercropping of *L. scindicus* and *C. ciliaris* give higher yield than their sole cropping.
	This system will be better for development and renovation of pastures and rangelands.
b) Pellets sowing	This method can be adopted for dry sowing as well as wet sowing to prevent loss of grass seed on windy days and from birds and ants.
	The pallets made by mixing in a particular proportion of 100-125 g seed: 3500 g clay: 250 g FYM: 250 g sand with a desired quantity of water and dried in shade for 24 hours with hand *chazlla or* a simple rotary pellet-making device developed by CAZRI, jodhpur [37].
	Suitable size of pellets should be 0.5 cm contains 2–3 seeds.

Package & Practices	Description
Sowing time, Seed rate and Spacing	Suitable sowing time as dry sowing is before the onset of monsoon under wet condition after the rains. The optimum seed rate is very important for getting the desired plant population in the field otherwise, growth of clumps at later stages are badly affected with heavy competition for moisture and nutrient as well as the space. 3–4 kg/ha seeds will be sufficient for one hectare area [35] and recommended crop geometry for sewan grass is 75–100 cm x 50–75 cm.
Fertilizer management	Desert soil has many advantages (better water releasing capacity) and disadvantages (poor water holding capacity) in terms of rainfed cropping of grasses as well as crops. Nitrogen, Phosphorus and Potassium uptake significantly higher in half-yearly cuttings than annual cuttings. Before sowing FYM or other compost including sheep and goat manures should be added approximately 5–7 t/ha and the recommended basal application is 30 kg nitrogen (two split doses) +40 kg P_2O_5 /ha has been found effective and economical dose for better establishment and higher forage yield [38]. The side placement has been found better than the broadcasting method of fertilizer application in the sandy soils.
Use of Bio regulators	Foliar spray of Thiourea (0.05%) and GA_3 (10 ppm) have positive significant effect on the seed yield of sewan grass. It induces the translocation of nutrient another part and many metabolic activities.
Irrigation management	Sewan is generally managed in natural rangelands in rainfed condition and major growing period is monsoon season. It is believed that if sewan is irrigated the productivity of grass will decrease and its life span will decrease form 10 years to 5 years. However, the experiment conducted at CAZRI, Jodhpur has shown that light irrigation through sprinklers with supplementation of nitrogen has increased yields (green forage yield 25.1 tones/ha and dry forage yield of 8.8 tones/ha) over the years [39]. It has been earlier reported that *L. scindicus* showed maximum water and energy use efficiency as compared to *C. ciliaris* and *C. setigerus*.
Irrigation scheduling	1. I cut at the end of August (rainfed), 2. I and II irrigations of 100 mm in October and November through sprinkler irrigation system with a cutting at the end of each month, 3. III irrigation of 100 mm at mid February and cutting at the end of March, and 4. IV irrigation of 100 mm at the end of March and cutting at the end of June.
Forage production	The highest green forage and dry matter yields recorded 88.57, 29.08 and 95.38, 30.19 q/ha from I and II cutting was recorded at 40 kg N/ha when applied full dose in July, respectively [38, 40].
Weed management	Weeds compete with the grass seedlings especially at the initial stages of its growth and development. Therefore, for efficient use of available soil moisture and nutrient by grasses the eradication of weeds is very important at the initial stage. It has been observed that although hand weeding is expensive but more effective than the chemical weeding. The grasses have fodder value hence chemical weeding is not advised. It has been proved that two weeding by hand hoe and after 20 days have been found effective and remunerative.

Package & Practices	Description
Harvesting	The sewan grass is ready to harvest after about 35 to 40 days after the effective rains. The nutritive status of fodder at maturity stage is lowest but total dry matter is more. Therefore, nutritive value of fodder yields should be harvested at green stage after the flowering. The harvesting can be done using mechanical harvester and grass cubes can be made for its unchaffed storage. In general, the grass is harvested by sickle and after drying, it is chaffed by the chaffing machine.
Ageing and productivity	Due to perennial nature, increase in age of the clumps increase green forage yield (15–17 q/ha) [41].
Soil Fertility	An experiment conducted at CSWRI, Bikaner from 2001 to 2003 revealed that the cultivation of sewan has non-significant effect on the soil EC and pH but soil organic carbon increase gradually. Available nitrogen, phosphorus and potassium significantly increased in the soil after three years [42].
Yield	From a well-managed sewan grass with good plant population 20–25 kg/ha seed and 35–40 q/ha dry forage can be produced. The yield potential is quite high *i.e.* about 250 kg/ha/year.

Table 4.
Environmental features and Agronomic practices for sewan grass production.

Alternate land use system	Remarks	References
Horti-Pastural system	Earlier study on this system has been revealed that sewan grass growth not affected with Horticultural intercrop (*Ziziphus mauritiana*)	[22]
Silvi-Pastural system	It has reported that in wastelands areas of arid regions, sewan grass intercropped with forest trees (*Acacia nilotica*, *Acacia tortilis*, *Acacia senegal* and *Colophospermum mopane*) and observed that sewan grass utilized moisture below 2 m soil surface whereas the trees takes the moisture more than 2 m depth. Thus, the system ensures best utilization of rainfall water and maintains temperature for biomass production. Some of the Silvi-Pastural systems with sewan grass are sewan + *C. mopane*, sewan + khejri, sewan + khejri + *C. mopane*, Sewan + *A. tortilis*, Sewan + *Acacia nilotica*, sewan + *Acacia senega*, sewan + *Ziziphus mauritiana*, *C. mopane* + *L. sindicus* + cowpea, *L. sindicus* + cowpea + *C. mopane* + *H. binnata*	[22–24]
Agri-Pastoral system		
Mixed cropping	Old practice among the farmers of western Rajasthan. This grass mixed with legumes increase fresh forage and dry matter yields. Example: *L. sindicus* + *C. ciliaris*, *L. sindicus* + *C. ciliaris* + *C. setigerus*, *L. sindicus* + *D. lablab*	[25, 26]
Intercropping	Intercropping of arid legumes (mungbean, moth and guar) with perennial grass include sewan, buffel grass and birdwood grass help to increase yields and stabilizes the economy of arid zone farmers.	[27]
Strip cropping system	Two crops of different growth habit are grown in a specified width of strips for enhancing the land productivity and reducing the soil erosion. Examples: Sewan + Mothbean (1:4), Sewan + Guar (1:4)	[28, 29]

Table 5.
Alternate land use system with Lasiurus sindicus.

and resource conservation in arid tract of Rajasthan. Sewan grass can be used as hay and silage during lean period when fodder is not available for livestock. The silage is a good quality fodder and can be fed to the animals during the off-season. An experiment conduced at CAZRI, Jodhpur and results revealed that quality silage of sewan fodder can be increased by adding 1–2% urea, 10% juggary and 4% starter culture (*Lactobacillus* culture) [41]. Sewan grass hay can be utilized as an excellent feed for dairy cattle that can be prepared by harvest at the proper physiological stage of growth and well cured to 20 per cent or less moisture.

An experiment conduced at CSWRI, Bikaner and results revealed that dry matter consumption through sewan hay was found to be higher than *C. ciliaris* hay where as digestibility of dry matter was lower in sewan hay than *C. ciliaris* [42].

7. Breeding efforts and achievements

Plant breeding deals with principles and procedures to improve the genetic constitution of crop species based on two basic principles such as creation of variation and selection. Naturally occurring variations in sewan grass already exist due to its cross-pollinated nature. Diversity existing among the germplasm help to select the diverse parents that help to introgression or combine the trait of interest into the elite cultivar [44] which can be estimated by clustering approaches such as Metroglyph analysis [45], D^2 statistics [46], Principal Component Analysis (PCA) [47] and molecular markers. Sewan grass is used for fodder purposes so that forage yield is the economically important complex or super trait. Direct selection for yields *per se* cannot be very effective. The study of inter-relationships is necessary for understanding the association of component traits with complex characters. Generally positive association between yield and component traits is beneficence for crop except maturity and anthesis traits in arid region otherwise it is advisable to break linkage drag between traits which made be possible through various population improvement strategies i.e. recurrent selection and its modifications, Disruptive selection mating and Marker-assisted recurrent selection, genomic selection, Genome editing methods etc. Now a day, in sewan grass, recurrent selection method and its modifications are being popularized for population improvement and varietal development. A lot of effort is still required to move applications for plant breeding beyond the experimental scale in sewan grass; however, Yadav and Krishna [48]; Shekhawat et al. [49]; Sanadya et al. [44, 50, 51]; have been screened large number of accessions of sewan grass for yield and its component traits and revealed that tillers number and dry matter yield are those characters showing high amount of variation and green fodder yield showed strong positively significant correlation with spike length, tillers number and dry matter yield. Sanadya et al. [50] have classified large number of sewan grass accessions into seven clusters using the Metroglyph method and Sharma et al. [11] grouped sewan grass accessions into five clusters using RAPD and ISSR markers. Chowdhury et al. [52, 53] found *nifH* gene in the rhizospheric region of sewan grass and also studied on diversity of 16sRNA and reported that sewan grass roots have been affiliated with a few of the nitrogen fixation bacteria *i.e. Pseudomonas pseudoalcaligenes*, *Azospirillum brasilense*, *Rhizobium* sp., and uncultured bacteria.

8. Conclusion

Arid zones are known to be fragile ecosystems in which various grasses have been introduced that tolerate high temperatures and low rainfall (below 250 mm) such as buffel-grass (*Cenchrus ciliaris*), birdwood grass (*Cenchrus setigerus*), marvel grass (*Dichanthium annulatum*), khavi grass (*Cymbopogon jawarancusa*), bermuda grass (*Cynodon dactylon*), wire grass (*Eleusine compressa*), sewan grass (*Lasiurus scindicus*), pan dropseed (*Sporobolus marginatus*), *tantia* (*Dactyloctenium sindicum*), halfa grass (*Desmostachya bipinnata*) etc. among them sewan grass is more popular because of good nutritive value and soil binder properties. This grass can be inter-cropped with other grasses, arid-legumes and desert trees with numerous alternate land use systems such as Agri-Pastoral system, Horti-Pastoral system, Silvi-Pastoral system and Agroforestry to complete, ecologically sustainable livelihood system. Although salt tolerance, drought tolerance, soil binder, nitrogen fixation, alternate land use system, ecofriendly nature, good palatilibity and high digestibility for livestock still plant breeders are not showing interest to popularize it to be farmers.

There are many reasons behind low popularization of sewan grass on farmer's field such as sewan grass found in extreme areas (high temperature), modern-ization in agriculture, cultivation of economically important crops, researches limited to agronomic perspective, limited R & D, seeds are very low weight and environmental conditions are highly variable (sandstroms), poor education and awareness, overgrazing, low profitable than economic important crops, uneven pod setting, non-synchronous maturation, present land utilization does not permit any more good land to be put for fodder production, and no governmental policies for conservation of sewan grass germplasm. Therefore, to meet both present and future demands, policies need to be supportive of the development of these tradi-tional Agroforestry systems. Sewan grass has numerous qualities such as lodging resistance, drought tolerance, C_4 grass, associated with beneficial bacterial colonies but still facing negligence from scientist communities. Therefore, these traits can be utilized for germplasm enhancement and it is mentioned earlier that intercropping of sewan with other crops and trees or shrubs also help to prevent soil erosion and maintain soil fertility. There is needed to be popularized fodder of sewan grass to the farmers so that can conserve sewan grass germplasm and generate extra income for their livelihood. If improvements could be made in forage quality, especially more high yield varieties with good nutrition, then potentially huge improvements in the animal production can be made. In conclusion, utilizing the information obtained from the research effort to improve grain crops and the knowledge gath-ered from model systems like *Brachypodium* and setaria, offers an excellent future perspective for improving the nutritional quality and yield for forage crops. The sustainable or ecological intensification of grass-based food production systems provides an opportunity to align the ever increasing global demand for food with the necessity to re-green ruminant production. Still integration of traditional breed-ing with modern approaches are missing in sewan grass therefore, modern genetics should be quickly integrated into the current conservation, use and improvement strategies to address nutritional quality and palatability concerns, in sewan grass.

Author details

Sanjay Kumar Sanadya*, Surendra Singh Shekhawat and Smrutishree Sahoo
Department of Genetics and Plant Breeding, Swami Keshwanand Rajasthan
Agricultural University, Bikaner, Rajasthan, India

*Address all correspondence to: sanjaypbg94@gmail.com

IntechOpen

References

[1] Bhagmal, Singh K, Roy A, Ahmed S, Malviya D. Forage crops and grasses. In: Handbook of Agriculture. sixth. ICAR, New Delhi; 2011. p. 1353-417.

[2] Anonymous. Grassland Index. A searchable catalogue of grass and forage legumes. FAO, Rome, Italy. 2010.

[3] Anonymous. Ecocrop database, Food and Agricultural Organization, Rome, Italy. 2010.

[4] Gupta RC, Chauhan HS, Saggoo M, Kaur N. Cytomorphology of some grasses (Poaceae) from Lahaul-Spiti (Himachal Pradesh), India. Biolife. 2014;2(4):1234-47.

[5] Quattrocchi U. CRC World Dictionary of Grasses: Common Names, Scientific Names, Eponyms, Synonyms and Etymology. CRC Press, Boca Raton, USA; 2006. 2408 p.

[6] El-Keblawy A, Ksiksi T, El Alqamy H. Camel grazing affects species diversity and community structure in the deserts of the UAE. J Arid Environ [Internet]. 2009;73(3):347-54. Available from: http://dx.doi.org/10.1016/j.jaridenv.2008.10.004

[7] Gohl B. Feed in the Tropics. 1982.

[8] Sahoo S, sharma A, Sanadya SK, Kumar A. Character association and path coefficient analysis in mothbean germplasm. Int J Curr Microbiol Appl Sci. 2018;7(8):833-9.

[9] Assaeed A. Estimation of biomass and utilization of three perennial range grasses in Saudi Arabia. J Arid Environ. 1997;36:103-11.

[10] Assaeed A, Al-Doss A. Seedling competition of Lasiurus scindicus and Rhazya stricta in response to water stress. J Arid Environ. 2001;49:315-20.

[11] Sharma R, Rajora M, Dadheech R, Bhatt R, Kalia R. Genetic diversity in sewan grass (Lasiurus sindicus Henr.) in the hot arid ecosystem of Thar Desert of Rajasthan, India. J Environ Biol. 2017;38:419-26.

[12] Khan MF, Anderson DM, Nutkani MI, Butt NM. Preliminary results from reseeding degraded Dera Ghazi Khan rangeland to improve small ruminant production in Pakistan. Small Rumin Res. 1999;32(1):43-9.

[13] Nagpal A, Kiradoo B, Purchit R, Mal G, Kumar R. Comparative studies on stall feeding and continuous pasture grazing systems on camel production. Indian J Anim Nutr. 1998;15:151-7.

[14] Nagpal A, Sahani M, Roy A. Effect of grazing sewan (Lasiurus sindicus Henr.) pasture on female camels in arid ecosystem. Indian J Anim Sci. 2000;70(9):968-71.

[15] Nagpal A, Saini N, Roy A, Sahani M. Nutrient utilization in camels fed sewan grass with or without ardu (Ailanthus excelsa) leaves. Indian J Anim Nutr. 2004;21:111-4.

[16] Thakur S, Mali P, Patnayak B. Evaluation of sewan pasture with or without supplementation of crushed clusterbean (*Cyamopsis tetragonoloba*). Indian J Anim Sci. 1985;55(8): 711-4.

[17] Bhati G, Mruthyunjaya. Economics of sheep farming on different pastures in arid land of western Rajasthan. Indian J Anim Sci. 1983;53(7):732-7.

[18] Gupta A, Joshi D. Effect of grazing on protein and mineral composition, and in vitro dry matter digestibility of different pasture species of the arid zone. Indian J Anim Sci. 1984;54(3):270-4.

[19] Rathore M. State level analysis of drought policies and impacts in Rajasthan, India. 2005. (6). Report No.: 93.

[20] Shankarnarayana K, Rao G, Ramanna B. Grassland Productivity and its associative climatic characteristics in western Rajasthan. Trop Ecol. 1985;26:157-63.

[21] Faroda A. Arid zone research - An overview. In: Faroda A, editor. Fifty Years of Arid Zone Research in India. Central Arid Zone Research Institute, Jodhpur; 1998. p. 1-16.

[22] Rao A, Singh K, Singh R. Climatic consideration in the development and management of Silvipasture system. In: Yadav M, Singh M, Sharma S, Tiwari J, Vurman U, editors. Silvi-pastoral system in arid and semi arid ecosystem. CAZRI, Jodhpur; 1995. p. 13-23.

[23] Anonymous. Annual Report. Central Sheep and wool Research Institute, Arid Campus, Bikaner. 1992.

[24] Dhir R. Characteristics and behavioral analysis of arid and semi arid areas soils. In: Yadav M, Singh M, Sharma S, Tiwari J, Vurman U, editors. Silvi-pastoral system in arid and semi arid ecosystem. CAZRI, Jodhpur; 1995. p. 178-9.

[25] Das R, Bhatt G. Compatibility trial on grass and legume at Jodhpur. Annual Report 1975, CAZRI, Jodhpur. 1975.

[26] Pal S, Bhati T. Forage performance of different types of sown and natural pastures. Annual Report 1980. 1980.

[27] Mudgal R, Patnayak B. Effect of different methods of sowing on establishment of sewan pasture in arid region of Rajasthan. Annual Report 1989. 1989.

[28] Misra, DK. Agronomic investigations in arid zone. In:

Proceedings of symposium on Problems of Indian arid Zone. Ministry of education, Govt. of India. New Delhi; 1971. p. 165-9.

[29] Singh A, Yadav R, Bawa A. Integrated land use system for management of Indian Arid zone. In: Proceedings of Indo-US workshop from January 9-14, 1984. Central Arid Zone Research Institute, Jodhpur; 1984. p. 64.

[30] Singh K. Grassland and pasture development in arid region of western Rajasthan. In: Kolarker A, Joshi D, Sharma K, editors. Rehabilitation of Degraded Arid Ecosystem. Scientific Publishers, Jodhpur; 1992. p. 121-6.

[31] Rajora M, Singh M, Jindal S. Seed harvesting in L. sindicus following hand cutting seed heads. Range Manag Agrofor. 2006;27(2):77-81.

[32] Lahiri A, Kharbanda B. Germination inhibitors in spikelets of glumes of L. Sindicus and C. ciliaris. Ann Arid Zone. 1963;1:114-26.

[33] Bawa A, Gupta I, Sharma B. Natural resource and their management in rangelands of western Rajasthan. Curr Agric. 1988;12:33-56.

[34] Singh M, Venkatesan K, Burman U. Enhancing Seed Set and Seed Yield of Sewan Grass (Lasiurus sindicus) through Physiological Approaches. In: Roy M, Malaviya D, Yadav V, Singh T, Sah R, Vijay D, et al., editors. 23rd International Grassland Congress. y Range Management Society of India; 2015.

[35] Yadav M. Potential grasses for arid zone. In: Singh A, Somani L, editors. Dry land resources and Technology. Scientific Publishers, Jodhpur; 1991. p. 115-22.

[36] Yadava N, Beniwal R. Effect of nitrogen on productivity of grasses in sole and intercropping system in arid

zone under rainfed condition. Indian J Agron. 2000;45(1):82-5.

[37] Yadav M. Pasture establishment Technique. In: Yadav M, Singh M, Sharma S, Tiwari J, Vurman U, editors. Silvipasture System in Arid and Semi Arid Ecosystem. Central Arid Zone Research Institute, Jodhpur; 1997. p. 193-5.

[38] Yadava N, Singh N, Soni M, Beniwal R. Response of sewan to saline water irrigation and fertilizer application for fodder production in arid western Rajasthan. Curr Agric. 2004;28(1-2):21-32.

[39] Singh S, Singh Y, Singh K. Water use and production potential of sprinkler irrigated Sewan pasture in Thar desert. Indian J Agric Sci. 1990;60(1):23-8.

[40] Mudgal R. Effect of nitrogen and cutting on sewan grass pasture under rainfed condition. Annual Report 1989-90. 1989.

[41] Anonymous. Ensiling of desert grasses. Annual Report 1990-91, Central Arid Zone Research Institute, Jodhpur. 1990.

[42] Gill S, Ratan R. Manipulation of soil fertility with crop residue lay farming and crop sequencing for eco-friendly higher production. In: Annual Report 2003-04. 2004.

[43] Narain P. Dryland Management in Arid Ecosystem. J Indian Soc Soil Sci. 2008;56(4):337-47.

[44] Sanadya SK, Shekhawat S, Sahoo S. Variability and inter-relationships of quantitative traits in Sewan grass (Lasiurus sindicus Henr.) accessions. Int J Chem Stud. 2018;6(6):1843-6.

[45] Anderson R. A semigraphical method for the analysis of complex problems. Proc Natl Acad Sci USA,. 1957;43:923-7.

[46] Mahalanobis P. A statistical Study at Chinese head measurement. J Asiat Soc Bengal. 1928;25:301-77.

[47] Pearson K. On Lines and Planes of Closest Fit to Systems of Points in Space. London, Edinburgh, Dublin Philos Mag J Sci. 1901;2(11):559-72.

[48] Yadav M, Krishna G. Studies on variability, correlations and path analysis in desert pasture grass Lasiurus sindicus Henr. Ann Arid Zone. 1986;25(2):157-63.

[49] Shekhawat S, Garg D, Pundhir P, Joshi P, Yadav N, Chhipa R. Evaluation of Lasiurus sindicus Henr. collection for their performance in zone IC of Rajasthan. Forage Res. 2003;29(2):76-8.

[50] Sanadya SK, Shekhawat S, Sahoo S, Kumar A, Kumari N. Metroglyph analysis in sewan grass (Lasiurus sindicus Henr.) accessions. Forage Res. 2018;44(2):86-9.

[51] Sanadya SK, Sahoo S, Baranda B, Sharma R. Study on correlation coefficient and path coefficient analysis in the accessions of Sewan grass (Lasiurus sindicus Henr.) for green fodder yield and related traits. J Pharmacogn Phytochem. 2019;(SP3):45-8.

[52] Chowdhury SP, Schmid M, Hartmann A, Tripathi AK. Diversity of 16S-rRNA and nifH genes derived from rhizosphere soil and roots of an endemic drought tolerant grass, Lasiurus sindicus. Eur J Soil Biol. 2009;45(1):114-22.

[53] Chowdhury SP, Schmid M, Hartmann A, Tripathi AK. Identification of diazotrophs in the culturable bacterial community associated with roots of Lasiurus sindicus, a perennial grass of Thar Desert, India. Microb Ecol. 2007;54(1):82-90.